ITM 1

SPRINGER SERIES IN PHOTONICS   5

Springer
Berlin
Heidelberg
New York
Hong Kong
London
Milan
Paris
Tokyo

Physics and Astronomy   ONLINE LIBRARY

http://www.springer.de/phys/

SPRINGER SERIES IN PHOTONICS

Series Editors: T. Kamiya   B. Monemar   H. Venghaus   Y. Yamamoto

The Springer Series in Photonics covers the entire field of photonics, including theory, experiment, and the technology of photonic devices. The books published in this series give a careful survey of the state-of-the-art in photonic science and technology for all the relevant classes of active and passive photonic components and materials. This series will appeal to researchers, engineers, and advanced students.

Series homepage – http://www.springer.de/phys/books/ssp/

Martin A. Green

Third Generation Photovoltaics

Advanced Solar Energy Conversion

With 63 Figures

 Springer

Professor Martin A. Green
University of New South Wales
Centre of Excellence for Advanced Silicon Photovoltaics and Photonics
Sydney, NSW, 2052, Australia

Series Editors:

Professor Takeshi Kamiya
Ministry of Education, Culture, Sports,
Science and Technology,
National Institution for Academic Degrees,
3-29-1 Otsuka, Bunkyo-ku,
Tokyo 112-0012, Japan

Dr. Herbert Venghaus
Heinrich-Hertz-Institut
für Nachrichtentechnik Berlin GmbH
Einsteinufer 37
10587 Berlin, Germany

Professor Bo Monemar
Department of Physics
and Measurement Technology
Materials Science Division
Linköping University
58183 Linköping, Sweden

Professor Yoshihisa Yamamoto
Stanford University
Edward L. Ginzton Laboratory
Stanford, CA 94305, USA

ISSN 1437-0379
ISBN 3-540-40137-7 Springer-Verlag Berlin Heidelberg New York

Cataloging-in-Publication Data applied for

Bibliographic information published by Die Deutsche Bibliothek Die Deutsche Bibliothek lists this publication in the Deutsche Nationalbibliografie; detailed bibliographic data is available in the Internet at <http://dnb.ddb.de>

Springer-Verlag Berlin Heidelberg New York
a part of Springer Science+Business Media

http://www.springer.de

© Springer-Verlag Berlin Heidelberg 2003
Printed in Germany

Data conversion: Marianne Schillinger-Dietrich, Berlin
Cover concept: eStudio Calamar Steinen
Cover production: *design & production* GmbH, Heidelberg

Printed on acid-free paper   57/3111   5 4 3 2 1

To Judy, Brie and Morgan

Preface

This text has its origins in my personal perceptions of how the photovoltaic industry is likely to develop as it expands to its full potential.

After my group's work in the early 1980s on improving silicon laboratory solar cell performance, we were able to capture most of this improvement in a commercially viable sequence, via the buried contact, laser grooved cell. Through the efforts of BP Solar, this has become one of the most successfully commercialised new cell technologies since then, with sales certain to exceed US$1 billion by 2010. This technology addresses the high material cost of "first-generation", silicon wafer-based photovoltaics by improving power out per unit investment in such material.

My group then was given the opportunity in the late 1980s to broaden its program into "second-generation" thin-film approaches, more directly addressing the issue of material costs. Although the then-favoured thin-film options, amorphous silicon, copper indium diselenide and cadmium telluride, all had their strengths, I believed there were quite fundamental limitations with each due to stability, resource availability and/or toxicity. Our previous success with silicon gave us confidence that we could develop a more desirable thin-film polycrystalline-silicon-on-glass technology almost from "scratch". A decade later, with rapidly increasing pilot-line module efficiencies being demonstrated by Pacific Solar with this "silicon on glass" approach, it became clear we had met our aim of developing a more viable thin-film option. This "second-generation" technology is capable of supporting the growth of the photovoltaic industry to beyond 2020, due to the quantum cost reduction it offers by eliminating wafers.

We then began to think about how this new technology might develop with time. Incremental refinements in material quality and device design were likely to increase efficiencies to close to 15%, comparable to the best presently with "first-generation" modules. We realised that, post-2020, with photovoltaics a large, profitable industry, there would be pressure to increase performance beyond this, since "second-generation" technology by then would be constrained by its own material costs. Just as the microelectronics industry relentlessly pushes towards smaller feature size to reduce costs, a mature photovoltaics industry would push towards ever-increasing conversion efficiency!

Tandem cells, where cells of different bandgap material are stacked on top of one another, offer a well-proved approach to increased efficiency. However, tandems involving compound semiconductors on top of thin-film silicon would not make a great deal of sense. There would be no compelling reason for using silicon in such a device, but rather compound material similar to that in the overlying device. Each cell added to a tandem stack also increases processing complexity and sensitivity to changes in the spectral content of sunlight. Were there alternative, more elegant approaches to increased performance, perhaps more compatible with the thin silicon on glass technology we had helped develop?

Both to answer this question and to meet the more mundane need to differentiate future research from work funded in the past, we were led to the concept of a third generation of photovoltaics. This would be differentiated from the two earlier generations by higher performance potential than from single junction devices. Other key criteria were that it use thin-films, for low material costs, and abundant, non-toxic materials. Although silicon is ideal in this regard, progress with molecularly based systems such as organic and dye sensitised cells and with nanostructural engineering in general, suggested other comparably attractive material systems may become available by 2020.

The first phase in our attempts to identify third generation candidates was to gain a clear understanding of the strengths and weaknesses of approaches suggested in the past for improving performance. It was also hoped that this re-examination might stimulate new ideas. This book documents the results of this phase. Taking a very broad view of photovoltaics, almost as broad as "*electricity from sunlight*", advanced photovoltaic options are analysed self-consistently with key features and challenges for successful implementation assessed. Although radiative inefficiencies readily can be incorporated, the main focus is on performance in the radiative limit. The rationale for this is that all successful photovoltaic devices must evolve towards this limit as argued above.

I would like to thank all who have stimulated my interest in photovoltaics since the early days, either by direct contact or by published work. I particularly thank those who took my postgraduate course on advanced photovoltaics during 2000, acting as guinea pigs for developing the text's first draft. Andrew Brown, Nils Harder and Holger Neuhaus deserve special mention for constantly challenging the material presented and for several graphs and tables in the text. I also thank Richard Corkish, Thorsten Trupke and Stuart Wenham and the high profile researchers on the Advisory Committee of the Centre formed to explore third generation options, particularly the longest serving members, Professors Antonio Luque, Hans Queisser and Peter Würfel. As the reader will note, the book also benefits from their past work. I also thank the Humboldt Foundation for a Senior Research Award and Professors Ernst Bücher, Ulrich Gösele and Rudolph Hezel for hosting associated visits during 2001 and 2002 where, amongst other activities, the manuscript was finalised. Finally, I thank Jenny Hansen for tireless efforts in producing diagrams plus many drafts of the text and Judy Green for support and companionship over the period this book was developed.

Bronte, Sydney   *Martin A. Green*
January, 2003

Table of Contents

1 Introduction

1.1 "Twenty-Twenty Vision"

Since the early days of terrestrial photovoltaics, many have thought that "first generation" silicon wafer-based solar cells eventually would be replaced by a "second generation" of inherently much less material intensive thin-film technology, probably also involving a different semiconductor. Historically, cadmium sulphide, amorphous silicon, copper indium diselenide, cadmium telluride and now thin-film silicon have been regarded as key thin-film candidates. Since any mature solar cell technology must evolve to the stage where cost is dominated by that of the constituent material, be it silicon wafers or glass sheet, it seems that high power output per unit area is the key to the lowest possible future manufacturing costs. Such an analysis makes it likely that photovoltaics, in its most mature form, will evolve to a "third generation" of high-efficiency, thin-film technology. By high-efficiency, what is meant is energy conversion values double or triple the 15-20% range presently targeted, closer to the thermodynamic limit upon solar conversion of 93%.

Tandem or stacked cells provide the best known example of how such high efficiency might be achieved. In this case, conversion efficiency can be increased merely by adding more cells of different bandgap to a stack, at the expense of increased complexity and spectral sensitivity. However, as opposed to this "serial" approach, better-integrated "parallel" approaches are possible that offer similar efficiency to even a stack involving an infinite number of such tandem cells. These alternatives will become increasingly feasible with the likely evolution of materials technology over the decades to 2020. This book discusses a range of these options systematically as well as paths to practical implementation. By clearly defining these options and identifying their strengths, weaknesses and areas where further work is required, their development may be accelerated.

1.2 The Three Generations

Most solar cells sold in 2003 were based on silicon wafers, so-called "first generation" technology (Fig. 1.1). As this technology has matured, its economics have become dominated increasingly by the costs of starting materials already

Fig. 1.1: Example of "first-generation" wafer-based technology (BP Solar Saturn Module using UNSW buried contact technology).

made in high volume and hence with little potential for cost reduction, such as silicon wafers, toughened low-iron glass cover sheet and other encapsulants. From experience with other technologies, this trend is expected to continue as the photovoltaic industry continues to mature. For example, a recent study (Bruton et al. 1997; Bruton 2002) of costs of manufacturing in a 500 MW/year production facility, about 10 times larger than the largest facilities in 2003, suggests material costs would account for over 70% of total manufacturing costs. The study therefore predicts lowest cost for high-efficiency processing sequences, provided these do not unduly complicate cell processing . Nonetheless, module efficiency above 16% is not contemplated even in this futuristic scenario.

For a prolonged period extending from the early 1980s, it has seemed that the photovoltaic industry has been on the verge of switching to a "second generation" of thin-film solar cell technology. Regardless of the semiconductor involved, thin-film technology offers prospects for a large reduction in material costs by eliminating the costs of the silicon wafer. Thin-film technology also offers other advantages, such as the increased size of the unit of manufacturing. This increases from the area of a silicon wafer ($\sim 100\,\mathrm{cm}^2$) to that of a glass sheet ($\sim 1\,\mathrm{m}^2$), about 100 times larger. On the efficiency front, with time, most would expect this "second generation" technology steadily to close the gap between its performance and that of "first generation" product.

As thin-film "second generation" technology matures, costs again will become progressively dominated by those of the constituent materials, in this case, the top cover sheet and other encapsulants (Woodcock et al. 1997). There will be a lower limit on such costs that, when combined with likely attainable cell efficiency (15% or 150 peak watts/m^2), determines the lower limit on photovoltaic module and, hence, electricity generation costs.

One approach to progress further is to increase conversion efficiency substantially. In principle, sunlight can be converted to electricity at an efficiency close to the Carnot limit of 95% for the sun modelled as a black-body at 6000 K and a 300 K cell (Chap. 2). This is in contrast to the upper limit, 31% on the same basis, upon the conversion efficiency of a single junction solar cell, as would limit silicon wafer and most present thin-film devices. This suggests the performance of solar cells could be improved 2-3 times if fundamentally different underlying concepts were used in their design, ultimately to produce a "third generation" of high performance, low-cost photovoltaic product.

There would be an enormous impact on economics if these new concepts could be implemented in thin-film form, making photovoltaics one of the cheapest known options for future energy production. Figure 1.2 shows this more graphically by showing possible production costs per unit area together with energy conversion efficiency ranges for the three generations of technology mentioned above. "First generation" wafer technology has high areal production costs and moderate efficiency, with few prospects for reducing the former below

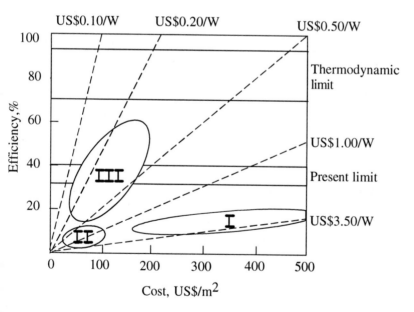

Fig. 1.2: Efficiency-cost trade-off for the three generations of solar cell technology; wafers, thin-films and advanced thin-films (year 2003 dollars).

about US$150/m^2 or the latter much above 20% (all dollar values in this book are year 2003 dollars). "Second generation" thin-film technology offers lower areal production costs, perhaps as low as US$30/m^2, but even more modest efficiency (presently in the 5-10% range). If the efficiency can be increased substantially using advanced "third generation" technology, much lower overall costs are possible even if there is a moderate increase in areal processing costs, relative to the second generation approaches.

1.3 Outline of Options

What are the prospects for developing thin-film cells based on new concepts capable of "third generation" performance? Fortunately, with the likely evolution of new materials technology over the coming decades, these appear quite good! Apart from tandem cells (Chap. 5), where efficiency can be increased progressively merely by stacking more cells on top of one another, a number of better-integrated "parallel" conversion approaches have also been suggested capable of similar efficiency, as previously mentioned.

One general class of approach that includes tandem cells is based on having *multiple energy threshold processes* available in the one device (Chap. 8). Other examples of this approach include multiple quantum well solar cells (Barnham and Duggan 1990) and devices based on the impurity photovoltaic effect (Green 1995). Recent work has clarified the limits on the efficiency of such devices by examining a general "three band" case (Luque and Marti 1997). In this case, excitation and recombination are allowed not only between the valence and conduction band, as in standard solar cells, but also between these bands and a third impurity band. Generalisation of this approach to more than three bands is possible, in principle, leading to *energy cascade cells* with potentially greatly improved performance.

A second class of approach is based on making fuller use of the energy of the high energy photons in sunlight such as by creating *multiple electron-hole pairs* per incident photon, as many as allowed by energy conservation (Chap. 7). Other quantum multiplication approaches are to create two lower energy photons from a single high energy photon, or an electron plus a lower energy photon.

A third class of approach is based on *hot carrier* effects (Chap. 6). Here, the sun's energy is stored in the vigorous motion of photoexcited carriers. To improve efficiency, these carriers must be collected before they get the chance to cool down to ambient temperature.

Another group of approaches is based on using sunlight to heat an absorber, with energy extracted from the heated absorber then being converted to electricity. Well known techniques of this type are the normal *solar thermal electric* approach, where heat from the absorber drives a heat engine, as well as *thermionics* and *thermoelectrics*. Another thermal approach is *thermo-*

photovoltaics, where the absorber re-radiates the absorbed energy. A solar cell is then used to convert the light emitted by this radiator (Coutts and Fitzgerald 1998). A more recent development is ***thermophotonics***, where the radiator is replaced by an electroluminescent device, such as a light emitting diode. The advantage is that the light intensity emitted by such a device can be much higher and much narrower in bandwidth for any given working temperature (Chap. 9).

In the following chapters, these and related concepts are reviewed and extended, with possible strategies for implementation explored, based on likely improvements in materials engineering over the next two decades. Although each approach mentioned above sounds fundamentally different from each of the others and from the tandem cell approach, common features emerge in their operation and performance limits. New concepts such as thermophotonic conversion are also introduced in this book, suggesting there is still scope for completely new approaches on which to base "third generation" devices, and which it is hoped this book may stimulate.

Exercise

1.1 Compare the total purchase costs of a nominally 1 kilowatt (peak) photovoltaic system for the following three choices of solar modules (at some stage in the future where the performance and cost figures mentioned have been demonstrated):

(a) "First generation" modules of 18% energy conversion efficiency at a projected cost of US$240/m^2;

(b) "Second generation" modules of 12% conversion efficiency at a projected cost of US$60/m^2;

(c) "Third generation" modules of 50% conversion efficiency at a projected cost of US$80/m^2.

Assume other area related costs total US$80/m^2 while non-area-related costs total US$1,000 for the system.

(Solar modules are normally given a rating under "peak" sunlight, corresponding to 1 kilowatt/m^2 intensity).

References

Barnham K and Duggan G (1990), A new approach to high-efficiency multi-band-gap solar cells, J Appl Phys 67; 3490-3493.

Bruton TM (2002), Music FM five years on: fantasy or reality, Conf. Record, Photovoltaics in Europe, Rome, October.

Bruton TM, Luthardt, G, Rasch K-D, Roy K, Dorrity IA, Garrard B, Teale L, Alonso J, Ugalde U, Declerq K, Nijs J, Szlufcik J, Rauber A, Wettling W and Vallera A (1997), A study of the manufacture at 500 MWp p.a. of crystalline silicon photovoltaic modules, Conf. Record, 14th European Photovoltaic Solar Energy Conference, Barcelona, June/July, pp. 11-16.

Coutts TJ and Fitzgerald MC (1998), Thermophotovoltaics, Scientific American, 68-73, September.

Green MA (1995), Silicon Solar Cells: Advanced Principles and Practice, (Bridge Printery, Sydney).

Luque A and Marti A (1997), Increasing the efficiency of ideal solar cells by photon induced transitions at intermediate levels, Physical Review Letters 78; 369.

Woodcock JM, Schade H, Maurus H, Dimmler B, Springer J and Ricaud A, (1997) A study of the upscaling of thin-film solar cell manufacture towards 500 MWp per annum, Conf. Record, 14th European Photovoltaic Solar Energy Conference, Barcelona, June/July, pp. 857-860.

2 Black-Bodies, White Suns

2.1 Introduction

A black-body is simply a body or object that is a perfect absorber of light and therefore, by a fundamental reciprocal relation, a perfect emitter. Although the perfect black-body represents a mathematical ideal, physical objects and devices can approach black-body properties reasonably closely. Even though defined so simply, black-bodies have played a surprisingly large role in the evolution of physics. Attempts to understand the spectral distribution of light emitted by heated black-bodies led directly to the development of quantum mechanics (Duck 2000). One particular approximation to a black-body, the paradoxically white and bright sun, has had an even more significant impact on human culture, being indispensable to human life itself.

Black-bodies feature prominently in the theory of the limiting performance of solar cells. Not only is the radiation emitted by the sun a good approximation to that from a very hot black-body, but an ideal solar cell would be expected to be a good absorber of light and hence be related to a black-body in some way. The details of this relationship forms the focus of many of the following chapters. An additional reason for the interest in black-bodies in solar cell theory is that, given their prominence in the development of physics, their thermodynamics have been thoroughly explored. This is particularly helpful when seeking to evaluate limiting solar cell performance.

2.2 Black-Body Radiation

A simple way to make a black-body is to put a small hole in an otherwise fully enclosed cavity (Fig. 2.1). Any ray entering the cavity via this hole will have great difficulty escaping, particularly if the cavity walls have moderately good absorption properties. The cavity therefore acts as an excellent absorber of radiation entering the hole. As a result, the area of the hole acts like an almost ideal black-body radiator for the inverse emission process.

From experimental data for such cavities, Max Planck first guessed the correct expression for the spectral variation of black-body radiation in 1900 and then attempted to derive this expression theoretically (Duck 2000; Bailyn 1994). Simply stated, he had to stretch both statistics and physics to achieve his goal! In

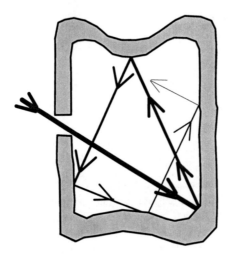

Fig. 2.1: A cavity with a small hole in it acts as an almost perfect absorber of light and therefore emits light from the hole with almost ideal black-body properties.

the latter area, he attempted to understand how his expression arose from the interaction of light with molecules of the black-body, modelled as oscillators. By working backwards to get the correct result, he was forced to hypothesise that these oscillators possessed energy only in multiples of a fundamental quantum.

Albert Einstein first saw the full ramifications of the quantum idea. In 1905, he suggested that light itself must have an intrinsic particle aspect (Duck 2000; Bailyn 1994). By assuming that an electromagnetic wave was not continuous but lumped into particles, he could explain existing experimental results from the photoelectric effect very simply (ejection of electrons from illuminated metals). This also allowed a different approach from Planck's to calculating the properties of a black-body (Bose 1924). As summarised by Richard Feynman (Feynman et al. 1965), the two approaches are equivalent. From one point of view, the radiation emitted by a black-body can be regarded as being in equilibrium with a large number of oscillators, one for each frequency component of light, with each oscillator in different excited states. From the other, the same property can be analysed in terms of Einstein's particles. The number of particles having a particular energy correspond to the state of excitation of the corresponding oscillator.

Satyendra Nath Bose was first to derive the black-body formula from the quantum particle point of view in 1924 (Feynman et al. 1965). Bose showed that the quantum particle hypothesis combined with the statistics of these particles were enough in themselves to give Planck's formula (Bailyn 1994). The particle hypothesis is that light consists of quanta of energy $E = hf$ and momentum $k = hf/c$, giving the following expression for the spatial components of this momentum:

$$k_x^2 + k_y^2 + k_z^2 = h^2 f^2 / c^2 \qquad (2.1)$$

Using a forerunner of Heisenberg's Uncertainty Principle of 1926 ($\Delta x \Delta k_x \geq h$), Bose imagined a 6-dimensional $xyzk_xk_yk_z$ phase space with the phase space volume divided into a mosaic of cells each corresponding to a separate state and each of volume:

$$h^3 = dx\, dy\, dz\, dk_x\, dk_y\, dk_z \qquad (2.2)$$

At fixed x, y, and z, the three dimensional $k_x\, k_y\, k_z$ sub-space is of interest. A spherical cell in the $k_x\, k_y\, k_z$ sub-space between radius k and $k + dk$ contains photons with approximately the same frequency, $f = ck/h$. The total phase volume $d\Omega$ associated with these photons is:

$$d\Omega = 4\pi k^2 dk \iiint dxdydz = 4\pi V(hf/c)^2 d(hf/c) \qquad (2.3)$$

where V is the physical volume of the cavity being considered. The number of states enclosed is:

$$\rho_f df = (g/h^3)d\Omega = g(4\pi V/c^3)f^2 df \qquad (2.4)$$

where the degeneracy factor, g equals 2. This is because light, considered as a planar transverse wave, needs two parameters to specify how it is orientated (i.e., has two possible polarisation states for each value of f) or has two possible spins, if considered as a particle.

Most readers, as for the author, probably find it easier to think in 3-dimensional rather than 6-dimensional space. Some feeling for the previous mathematics can be gained by thinking in the 3-dimensional xk_xk_y space of Fig. 2.2, as might be important for fixed y, z and k_z. The k_xk_y plane is the analogue of the 3-dimensional $k_xk_yk_z$ space, in this example. The analogue of the spherical shell is the annular ring shown for three such planes in Fig. 2.2. The volume corresponding to k_x and k_y values within this ring within the 3-dimensional space is just the area of the ring multiplied by the extent of the region in the x-direction, for the fixed values of y and z. This corresponds to the tubular volume shown. The calculation is simplified since the geometry in the k_xk_y plane is independent of x-value. For the 6-dimensional case, the volume of the shell in the $k_xk_yk_z$ sub-space is independent of x,y and z co-ordinates. The volume in the 6-dimensional space is found for an incremental physical volume around any particular x, y, z co-ordinate and integrated over the entire physical volume. The latter converts to a multiplication due to the non-dependence upon spatial co-ordinates.

Bose then calculated the most probable distribution of photons amongst his mosaic of cells within phase space. He was able to show that the average number of particles is given by the function that now bears his name (together with

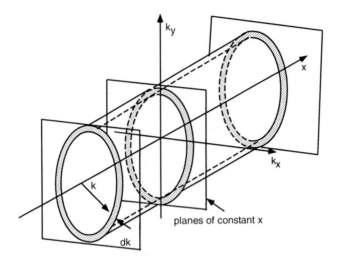

Fig. 2.2: 3-dimensional xk_xk_y space analogue of the calculation conducted in 6-dimensional $xyzk_xk_yk_z$ space in text.

Einstein's who was quick to realise the significance of Bose's work and to develop it). The Bose-Einstein distribution function is given by:

$$f_{BE} = 1/(e^{hf/kT} - 1)$$ (2.5)

The total photon energy per unit volume in the cavity in the frequency range df is therefore given by:

$$u_V \, df = \frac{8\pi (hf^3/c^3) df}{e^{hf/kT} - 1}$$ (2.6)

2.3 Black-Body in a Cavity

If a black-body is inserted in the cavity of Fig. 2.1, as in Fig. 2.3, several properties of the black-body can be deduced (Siegal and Howell 1992).

A black-body must be a perfect emitter of light as well as a perfect absorber. This follows since the black-body absorbs all incident radiation from the cavity. After a period of time, the black-body and the cavity will reach a common equilibrium temperature. Since there can be no net energy transfer when at a common temperature, the black-body must be emitting the maximum amount of radiation. This must follow because anything less than a perfect absorber would have to emit less to remain in equilibrium.

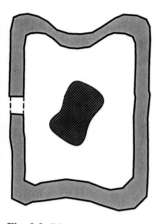

Fig. 2.3: Black-body in a cavity.

The radiation within the cavity must be isotropic. If the black-body, after equilibriating, were rotated or moved to another position in the cavity, it must remain at the same temperature since there are no heat inputs to the system. Consequently, it would emit the same radiation as before with the reasonable assumption that the emission rate is a function of temperature only. It follows that the black-body must be absorbing the same amount of radiation. For arbitrary geometries, this could only occur if the radiation were isotropic. By developing this argument, it can be shown that a black-body must be a perfect emitter in each direction and at each wavelength (Siegal and Howell 1992).

2.4 Angular Dependence of Emitted Radiation

Figure 2.4 shows the segment, dA, of a black-body surface emitting radiation in a direction defined by the angles θ and φ. Also shown is the projected area, dA_p of this element normal to the (θ, φ) direction, with an area of $dA \cos\theta$.

It is straightforward to show that the radiation flux emitted per unit solid angle and per unit projected area by a black-body must be constant. (This can be proved by imagining that the hemispherical shell shown in Fig. 2.4 represents part of the surface of a spherical black-body cavity exchanging radiation with the central element). As opposed to this, the radiation emitted per unit area of emitting surface must be proportional to $\cos\theta$. Surfaces with this characteristic are known as Lambertian, and feature prominently in solar cell light trapping theory (Green 1995). Figures 2.5(a) and (b) show this difference in the two emission characteristics, diagramatically.

The amount of radiation emitted into a finite range of angles (finite solid angle) also can be calculated using the construction of Fig. 2.4. A solid angle is defined as the area intercepted by the section of space that encompasses the solid

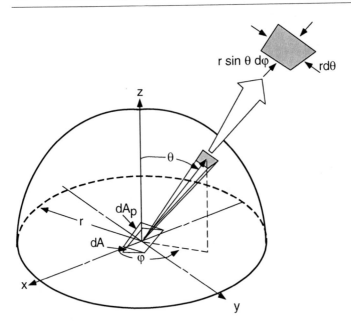

Fig. 2.4: Construction for examining angular dependence of emission from a small area, *dA*, of the surface of a black-body (after Siegal and Howell 1992).

angle, on an arbitrarily sized hemisphere centred at the angle's vertex, divided by radius of this hemisphere squared. Hence, the solid angle corresponding to the segment on the surface of the hemisphere of Fig. 2.4 defined by small variations $d\theta$ and $d\varphi$ equals $\sin\theta \, d\theta d\varphi$. The radiation emitted per unit solid angle and per unit projected area is R, independent of θ and φ for a black-body (regardless of whether "radiation" refers to energy, entropy or photon fluxes, or to one wavelength or to the radiation integrated over all).

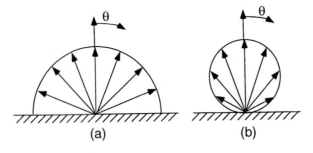

Fig. 2.5: Directional emission characteristics of a black-body (or Lambertian) surface: (a) Emitted flux per unit solid angle and per unit projected area normal to the emission direction; (b) per unit surface area.

Hence, the radiation passing through the element on the hemisphere's surface per unit emitting surface area, dA, equals $R\cos\theta\ sin\theta d\theta d\varphi$. Integrating over a range of θ and φ gives:

$$\int_{\varphi_1}^{\varphi_2}\int_{\theta_1}^{\theta_2} R\cos\theta\sin\theta\ d\theta\ d\varphi = R\int_{\varphi_1}^{\varphi_2}\int_{\sin\theta_1}^{\sin\theta_2}\sin\theta\ d(\sin\theta)d\varphi$$

$$= R\left(\frac{\sin^2\theta_2-\sin^2\theta_1}{2}\right)(\varphi_2-\varphi_1)$$

(2.7)

If $\varphi_1 = \theta_1 = 0$, $\theta_2 = \pi/2$ and $\varphi_2 = 2\pi$, the total radiation emitted per unit surface area passing through the hemisphere is calculated as πR.

Figure 2.4 can also be used to find the total light emitted by an elemental black-body surface area on Earth intercepted by the sun. Consider the case when the sun is directly overhead. In this case, $\sin\theta_2 = r_s/d_{es}$ and $\theta_2 = 2\pi$ where r_s is the radius of the sun (695,990 km) and d_{es} is the distance from the elemental area to the sun [calculable as the mean sun-earth distance, a mere 149,597,871 km, minus a small distance approximately equal to up to the Earth's radius, 6,378 km, divided by the air mass (Green 1982) corresponding to the sun's position]. Hence, $\theta_2 = 0.26657(1)°$ and the light intercepted equals $f_\omega\pi$ where f_ω equals $(r_s/d_{es})^2$, or $2.1646(1)$ x 10^{-5}. The number in brackets represents the uncertainty in the last digit of the preceding number due to changes in the position on Earth relative to the sun. Greater variability is introduced by the eccentricity of the Earth's orbit (1.67%), which introduces a similar percentage variability into θ_2 and about double this percentage into f.

If the element and sun were at the same temperature, this flow would have to be balanced by an equal and opposite flow from the sun. Hence, the same expression gives the total amount of the radiation emitted over the entire sun's surface that strikes an elemental area on Earth, where R is the corresponding radiation emitted at the sun's temperature. This result could be derived more directly by considering a huge sphere centred at the sun, but intersecting the Earth at the elemental area's position.

The fact that direct sunlight arrives over such a small range of angles can be used to advantage in photovoltaics. This compactness allows the direct component of sunlight to be concentrated, which increases cell efficiency by increasing cell voltage outputs. Alternatively, the small angular spread can allow the net rate of recombination within the cell to be suppressed by restricting the cell's angular acceptance of light and hence its total light emission (Green 1995; Araujo 1990). Both approaches give rise to the same limiting efficiency.

With free space between sun and the cell, the best that can be done for concentrated sunlight is to design an optical system that steers all the black-body radiation emitted by the incremental earth-based element onto the sun. In this case, all light emitted by the element into the hemisphere would reach the sun, i.e., an amount equal to πR (i.e., $f = 1$). By the same argument as before, a corresponding amount from the sun's surface would reach the incremental area with such optics. Hence, the maximum possible concentration level possible equals $(d_{es}/r_s)^2$ or 46,198(2), if the intervening medium is vacuum. (Other values for this quantity in the literature stem from either using rounded values for the different parameters involved or a less accurate solid-angle argument to deduce the amount of sunlight intercepted). At this concentration level, the intensity at the sun's surface is reproduced at the cell surface, a frightening thought but something that is close to achievable.

If the receiving element is immersed in a medium of refractive index, n, maximally achievable concentration is increased to $(d_{es}/r_s)^2 n^2$ (Smestad et al. 1990). Concentrators approaching this limiting performance actually have been built. The present record appears to be a concentration ratio of 84,000 corresponding to an output power of 7.2 kW/cm^2 (Cooke at al. 1990). Figure 2.6 shows the experimental set-up. Note that the refractive index of air under standard conditions is a little higher than unity (1.000278) but will be taken as unity throughout this text.

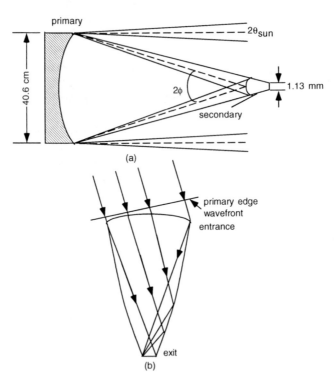

Fig. 2.6: Experimental high-performance concentrator consisting of (a) a parabolic mirror primary and (b) a compound parabolic secondary (after Cooke et al. 1990).

2.5 Direct and Diffuse Efficiencies

As opposed to the directional radiation in space, the Earth's atmosphere scatters incoming sunlight so that some is incident on terrestrial cells from all accessible directions in the sky. A terrestrial system that converts only the direct component of sunlight can therefore waste a lot of energy, even if it converts the direct very efficiently. Usually, the quoted efficiency for concentrating photovoltaic and solar thermal electric systems is based on only the direct component of incident sunlight, rather than the combined direct plus diffuse. Tracking of the sun is also essential once a reasonable concentration level is reached.

A standard non-concentrating photovoltaic system converts both direct and diffuse sunlight. Since sun tracking is not essential, such tracking is usually regarded as a bonus when implemented, giving up to 40% more energy compared to a stationary system of the same peak rating.

The overall efficiency of conversion, η, could therefore be regarded as a composite of a direct and diffuse component:

$$\eta = f_{dir}\eta_{dir} + (1 - f_{dir})\,\eta_{diff} \tag{2.8}$$

where f_{dir} is the fraction of available light that is direct, η_{dir} is the conversion efficiency for this light while η_{diff} is that for diffuse light. For a normal concentrator system, $\eta_{diff} = 0$, while for a normal non-concentrating system, $\eta \approx \eta_{diff}$. Fundamentally, $\eta_{diff} < \eta_{dir}$ due to the performance gains possible by taking advantage of light collimation. In both common cases, the overall η is appreciably less than that possible if all light were direct.

It is possible to improve on these norms. For example, the system shown in Fig. 2.7 could be designed to convert the direct component of sunlight with high efficiency while also efficiently converting the diffuse (Goetzberger 1990). The approach also accommodates losses due to non-idealities in the concentrating optics. (The diffuse converters would have to be very inexpensive to make such a scheme attractive in practice, however).

In the following chapters, both η_{dir} and η_{diff} will be calculated. The rationale for calculating the former is that it is the relevant efficiency, in principle, for space systems and for the conversion of the direct component of terrestrial sunlight. It also represents the highest value obtainable for any given conversion approach, allowing unambiguous comparison between different approaches and comparison also with thermodynamic limits.

Optimal η_{dir} is obtained when all light emitted by the cell strikes the sun. This can be ensured, in principle, by appropriate concentrating optics which produces f values of unity for both radiation emitted by the sun and by the cell. Alternatively, the cell can be designed, in principle, so that it absorbs only light from the small range of incident angles involved (Green 1995; Araujo 1990) and hence emits light only into this range (angularly selective black-body properties). In this case, f

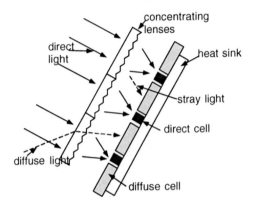

concentrating lenses

direct light

heat sink

stray light

direct cell

diffuse light

diffuse cell

Fig. 2.7: A combined direct-diffuse light converter capable of higher conversion efficiencies based on the total incident light intensity.

has the much lower value of f_ω (2.1646 x 10⁻⁵) for both sun and cell. Similar mathematics apply to both cases, although either approach is difficult to implement in practice. Concentration levels of 1,000 normally are regarded as very high for experimental photovoltaic systems, while there is not yet any demonstration of performance gain by the second approach. Intermediate options are possible whereby the sun is concentrated to less than the maximally possible concentration level with the cell then designed to accept only the resulting angular spread (Green 1995).

η_{diff} is calculated by applying the lower value of f for light received from the sun (f_ω), while applying the higher value of unity to that emitted by the cell.

2.6 Black-Body Emission Properties

From Sect. 2.2, expressions for the emitted black-body flux can be deduced regardless of whether particle, energy or entropy fluxes are being discussed. The flux of an extensive quantity carried by particles (an example of an extensive quantity is the total photon energy in the cavity), in the general case, can be related to its volume density by multiplying by a quarter of the mean particle velocity. In the present case, this means that the cavity quantities of Sect. 2.2 need to be multiplied by $c/4$ to convert to fluxes (see box below).

The particle, energy and entropy fluxes emitted by a black-body per unit surface area into a hemisphere over the energy range, E_1 to E_2 are given by:

$$\dot{N}(E_1, E_2) = \frac{2\pi}{h^3 c^2} \int_{E_1}^{E_2} \frac{E^2 dE}{e^{E/kT} - 1} \qquad (2.9)$$

$$\dot{E}(E_1, E_2) = \frac{2\pi}{h^3 c^2} \int_{E_1}^{E_2} \frac{E^3 dE}{e^{E/kT} - 1} \qquad (2.10)$$

$$\dot{S}(E_1, E_2) = \frac{\dot{E}}{T} - \frac{2\pi k}{h^3 c^2} \int_{E_1}^{E_2} E^2 \ln(1 - e^{-E/kT}) dE \qquad (2.11)$$

$$= \frac{4\dot{E}}{3T} - \frac{2\pi k}{3h^3 c^2} \left| E^3 \ln(1 - e^{-E/kT}) \right|_{E_1}^{E_2} \qquad (2.12)$$

Relationship Between Extensive Variables and Fluxes

To convert from quantities in a cavity such as energy per unit volume to fluxes (energy per unit area per second), it is apparent from dimensional analysis that the former have to be multiplied by a velocity. But what velocity?

If all light were moving in a direction perpendicular to the area dA of interest, the appropriate velocity would be $c/2$. This is because half the light in a volume cdA would go in each direction in unit time. However, within the black-body cavity, the radiation is isotropic and moves with velocity c in all directions. A fraction $d\Omega/4\pi$ can be considered going in any direction specified to within an element of solid angle $d\Omega$.

Consider a hole of area dA in the cavity as in Fig. 2.4 but, this time, assume the hemisphere penetrates into the cavity and that it represents a shell of finite thickness, dr. The area dA subtends a solid angle $dA\cos\theta/r^2$ from the point of view of the small volume $r^2\sin\theta d\theta d\varphi dr$. A fraction of the energy in this volume determined by this solid angle reaches the element dA in a time interval between t and $t + dt$, where $dt = dr/c$.

The amount of energy flowing through this hole in the time dt from this small volume equals $u_\nu c \cos\theta\sin\theta d\theta d\varphi dt dA$. The total amount arriving per unit area per unit time is given by:

$$\int_0^{\pi/2\pi} \int_0^{2\pi} u_\nu \, c\cos\theta \, \sin\theta \, \frac{d\theta d\varphi}{4\pi} = \frac{u_\nu c}{4}$$

This is the total emitted flux per unit time per unit area. For particles other than photons, a similar calculation shows that the volume density of an extensive variable is converted to a flux by multiplying by $\overline{v}/4$, where $\overline{v}$ is the average particle velocity.

Most readers may not be as familiar with entropy fluxes as with particle and energy fluxes. Chapter 3 will give a fuller discussion of the entropy flux and its importance in determining limiting efficiency. An interesting feature of the entropy expression is noted in passing. If we regarded the energy flux radiated by the black-body as being supplied from a heat reservoir at very close to the same temperature (similar to the way that the hydrogen fusion reaction at the sun's core supplies the energy emitted by the sun), the entropy flow from this reservoir to the area involved is given by the energy flux divided by T, different from the entropy flux emitted.

The above implies there is entropy production at the surface emitting the light (Planck 1959; De Vos and Pauwels 1983) to provide the entropy balance [the production rate is equal to the second term on the right of Eq. (2.11)]. Similarly, entropy generation is associated with light absorption. When calculated separately, the latter can give a negative contribution, although the overall balance is always be positive, becoming zero only when the emitted and absorbed light is identical.

As shown in Appendix C, the integrals involved in the above equations can be evaluated in terms of the standard Bose - Einstein integrals:

$$\beta_j(\eta) = \frac{1}{\Gamma(j+1)} \int_0^\infty \frac{\varepsilon^j d\varepsilon}{e^{\varepsilon - \eta} - 1} \tag{2.13}$$

where $\Gamma(j + 1)$ equals the Gamma function given by $j!$ for positive integers as arguments. For the case where $E_1 = 0$ and $E_2 = \infty$, the integral involved can be expressed in terms of $\beta_j(0)$ which equals the Reiman zeta function given by:

$$\xi(j+1) = \sum_{n=1}^{\infty} (1/n)^{j+1} \tag{2.14}$$

This sum can be expressed in terms of π for even integral arguments:

$$\xi(2) = \pi^2/6, \quad \xi(4) = \pi^4/90, \quad \xi(6) = \pi^6/945$$

Using these results gives:

$$\dot{N}(0,\infty) = \frac{2\pi(kT)^3 \Gamma(3)\xi(3)}{h^3 c^2} = \dot{E}(0,\infty)/2.70117kT \tag{2.15}$$

$$\dot{E}(0,\infty) = \frac{2\pi(kT)^4}{h^3 c^2} \Gamma(4)\xi(4) = \sigma T^4 \tag{2.16}$$

$$\dot{S}(0,\infty) = \frac{4\dot{E}(0,\infty)}{3T} = \frac{4\sigma T^3}{3} \tag{2.17}$$

where σ is the Stefan - Boltzmann constant given by:

$$\sigma = 2\pi^5 k^4/(15h^3 c^3) = 5.670400(40) \times 10^{-12} W/cm^2/K^4 \tag{2.18}$$

Note that the average photon carries the energy $2.70117kT$ [equal to $3\xi(4)/\xi(3)$ times kT].

If properties are less than ideal, these quantities can be multiplied by appropriate averaged emissivities (equal to the corresponding absorptivities). Wavelength dependent emissivities could be included within the integrals of Eqs. (2.9) to (2.11). Angularly dependent ones would need to be incorporated prior to this stage of development.

Exercise

2.1 Assuming the sun can be modelled as an ideal black-body at 6000 K, calculate the Earth's surface temperature, assuming its temperature is uniform over its entire surface and that the sun is the only source providing energy to it. Assume the following models for the Earth's radiative properties:

(a) Ideal black-body properties for the Earth;

(b) Grey-body properties with an absorptance of 0.7;

(c) Spectrally sensitive absorptance such that the absorptance averaged over the sun's energy spectrum is 0.7. The reduced emissivity at longer wavelengths, due to greenhouse gases, is modelled as a decreased absorptance of 0.6 averaged over wavelengths corresponding to the Earth's radiative emission.

(d) How much change in the latter absorptance is required to heat the Earth's temperature by 2°C?

References

Araujo GL (1990), Limits to efficiency of single and multiple bandgap solar cells, in Physical Limitations to Photovoltaic Energy Conversion (A. Luque and G.L. Araujo (eds.)), Adam Hilger, Bristol.

Bailyn M (1994), A Survey of Thermodynamics, AIP Press, New York, 530-545.

Bose SN (1924), Plancks Gesetz und Lichtquantenhypothese, Zeitschrift für Physic 26: 178-181.

Cooke D, Gleckman P, Krebs H, O'Gallagher J, Sage D and Winston R (1990), Sunlight brighter than the sun, Nature 346: 802.

De Vos A and Pauwels H (1983), Comment on a thermodynamical paradox presented by P. Würfel, J Phys C : Solid State Phys 16: 6897-6909.

De Vos A (1992), Endoreversible thermodynamics of solar energy conversion, Oxford University Press, Oxford.

Duck I, Sudarshan ECG (2000), 100 years of Planck's quantum, World Scientific, Singapore.

Feynman RP, Leighton RB and Sands M (1965), The Feynman lectures on physics: Volume III: quantum mechanics, Addison Wesley, Reading, MA, Chap. 4, 9.

Goetzberger A (1990), private communication.

Green MA (1982), Solar Cells: Operating Principles, Technology and System Applications, (Prentice-Hall, New Jersey). (Reprinted by Centre for Photovoltaic Engineering, University of New South Wales, available from author).

Green MA (1995), Silicon Solar Cells: Advanced Principles and Practice, (Bridge Printery, Sydney). (Available from author).

Landsberg PT (1983), Some maximal thermodynamic efficiencies for the conversion of black-body radiation", J Appl Phys 54 (1959): 2841-2843.

Landsberg PT and Tonge G, (1980), Thermodynamic energy conversion efficiencies, J Appl Phys 51: R1-R20.

Planck M (1959), The Theory of Heat Radiation, Dover, New York, [English translation of Planck M (1913), Vorlesungen über die Theorie der Warmestrahlung, Leipzig: Barth].

Siegel R and Howell JR (1992), Thermal Radiation Heat Transfer, 3rd ed., Hemisphere Publishing Corporation, Washington.

Smestad G, Ries H, Winston R and Yablonovitch E (1990), The thermodynamic limits of light concentrators, Sol En Matls 21: 99-111.

Stackel J (ed.) (1998), Einstein's Miraculous Year: Five Papers that Changed the Face of Physics, Princeton University Press, Princeton, 177-198.

3 Energy, Entropy and Efficiency

3.1 Introduction

Photovoltaics usually is associated with quite modest sunlight to electricity conversion efficiencies. Commercial "first generation" silicon wafer-based solar modules usually have efficiencies in the 10-15% range and "second generation" thin-films have even lower values of 4-9% (Schmela 2002). However, the thermodynamic arguments now to be outlined show that it is possible to conceive of solid-state conversion of sunlight to electricity at close to the Carnot efficiency of 95%. This value is the limit for the conversion of heat from a source at 6000K (the sun's photosphere) when combined with a sink temperature of 300K (the terrestrial ambient).

3.2 Energy and Entropy Conservation

The first and second laws of thermodynamics were succinctly stated by Rudolf Clausius in 1865: *"The total energy of the universe is constant. The total entropy of the universe strives to reach a maximum"* (Clausius 1865). By associating non-negative entropy production with any energy exchange, both laws can be expressed as conservation laws.

Energy is a reasonably familiar concept, with the following description particularly appropriate for photovoltaics. Energy is: *"The capacity for doing work. The various forms of energy ... include **heat, chemical, nuclear** and **radiant energy**. Interconversion between these forms of energy can occur only in the presence of **matter**. Energy can only exist in the absence of matter in the form of radiant energy"*. Matter itself is: *"A specialised form of **energy** which has the attributes of **mass** and extension in **space** and time"* (Lurarov and Chapman 1971).

Entropy is undoubtedly a less intuitive concept, but is physically associated with disorder or "spread" (Dugdale 1996). The less disorder or the smaller spread, the smaller the entropy. Clausius showed that entropy could be expressed as heat divided by temperature. The transfer of a small amount of heat therefore causes a larger change in entropy in a cold body than in a hot body, consistent with heat flowing from hot to cold.

In solar conversion, we are interested in nominally steady-state conditions and the equilibrium between energy fluxes, rather than incremental transfers.

Associated with an energy transfer as heat at rate $\dot{E}$ to or from a body at temperature T, is an entropy flux $\dot{E}/T$, if the transfer occurs in the presence of an infinitesimally small temperature differential. There is an additional entropy generation term if temperature gradients are involved, such as when there is imperfect thermal contact with a heat sink (Dugdale 1996).

Radiative energy transfer between bodies at nearly the same temperature is associated with an entropy flux given by the same expression. This is probably not too surprising since energy transfer, in this case, involves the exchange of photons, while heat transfer commonly involves phonons. Both are bosons, with similar fundamental properties.

One difference is that radiative energy transfer often involves energy transfer directly between bodies at vastly different temperatures. This leads to entropy generation, unlike the case of transfer between nearly equal temperatures, and to a different approach to treating corresponding energy and entropy fluxes. Associated with the radiative surfaces are an outgoing and incoming flux. Outgoing fluxes are those that would be emitted into a zero temperature environment. Incoming fluxes are those received from surrounding surfaces, as calculated from their emission into the zero temperature environment. This differs from the normal formulation of heat transfer, which is usually in terms of nett fluxes, with it not being usual to partition between incoming and outgoing fluxes in the above way.

3.3 Carnot Efficiency

The most general efficiency limit for photovoltaics is the Carnot limit, that applies to the system defined in Fig. 3.1.

Inputs are E_S, the heat energy flux supplied from the sun's interior to fuel its radiative emission and S_S, the corresponding entropy flux, given by E_S/T_S where T_S is the temperature of the sun's photosphere (assumed to be 6000K in future calculations). Outputs are an energy flux in the form of useful work $\dot{W}$, with zero associated entropy flux, and a heat flux $\dot{Q}$ rejected to the ambient. This has an associated entropy flux $\dot{Q}/T_A$ where T_A is the ambient temperature (300K in future calculations). Another input is an entropy generation flux, $\dot{S}_G$, associated with the conversion process, which must be non-negative and will be positive in any practical process, and sometimes a very large positive, depending on the details of the converter.

Including this term allows the first and second laws of thermodynamics to be expressed as an energy and entropy flux balance, respectively:

$$\dot{E}_S = \dot{W} + \dot{Q} \tag{3.1}$$

$$\dot{S}_S + \dot{S}_G = \dot{Q}/T_A \tag{3.2}$$

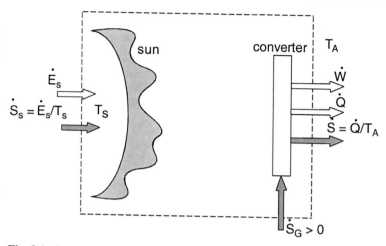

Fig. 3.1: System considered for calculating Carnot efficiency.

These equations are linked by the term in $\dot{Q}$. Combining (3.1) and (3.2) by eliminating this term gives:

$$\dot{E}_S = \dot{W} + T_A(\dot{S}_s + \dot{S}_G)$$ (3.3)

The conversion efficiency is then given by:

$$\eta = \dot{W} / \dot{E}_S = (1 - T_A / T_S) - T_A \dot{S}_G / \dot{E}_S$$ (3.4)

This has its maximum value of 95% when $\dot{S}_G$ has its minimum value of zero, corresponding to the Carnot efficiency for the conversion of heat energy supplied to the sun's photosphere to terrestrial electricity, or other useful work. An interesting feature of this derivation is that no information is required about the converter. The question naturally arises as to whether there is any converter that could, at least in principle, achieve this limiting Carnot efficiency.

The main requirement is that there be no entropy generation during the transmission, absorption or conversion of the sunlight, quite a difficult requirement to satisfy, even conceptually. Planck showed almost a century ago (Planck 1959), as also discussed more recently (De Vos and Pawels 1983), that energy transfer between two black-bodies involves unavoidable entropy production, unless both are at the same temperature. This argument can be generalised to finite entropy production in an absorber unless the absorber emits light of the same intensity as the sun at each wavelength. The problem is that there would be no net energy transfer under these conditions. It follows that, to achieve the Carnot limit, only infinitesimally small amounts of work could be produced, with recycling of most of the sun's energy.

Although this argument implicitly assumes reciprocity between light absorption and emission, an issue explored more fully in a later section, the conclusion is valid even when the possibility of non-reciprocal systems is taken into account. The result is not very satisfactory since, although efficiently using the sun's energy reaching the earth could have a big impact here, recycling to the sun some fraction of the twenty millionth of its output reaching the earth is not likely to generate the same enthusiasm there.

3.4 Landsberg Limit

To address the problem we really want to solve, the more restricted system shown in Fig. 3.2 needs to be considered. Inputs are the radiant energy and associated entropy fluxes from the sun plus the unavoidable entropy generation during the absorption and conversion processes. Outputs are again work and heat fluxes, plus the entropy flux associated with the latter. Additional outputs in this case are the energy re-radiated by the converter and the associated entropy flux.

The analysis proceeds as above, except the additional outputs are added to the right-hand sides of Eq. (3.1) and (3.2). This gives:

$$\eta = \dot{W} / \dot{E}_S = (1 - T_A \dot{S}_S / \dot{E}_S) - (1 - T_A \dot{S}_C / \dot{E}_C) \dot{E}_C / \dot{E}_S - T_A \dot{S}_G / \dot{E}_S \qquad (3.5)$$

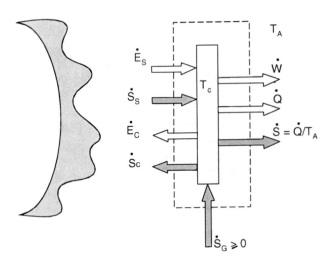

Fig. 3.2: System for calculating Landsberg efficiency limit.

Since we are now dealing with total radiant energy from the sun, rather than a nett heat flux, the entropy flux increases to $\frac{4}{3}\dot{E}_S/T_S$, as discussed in connection with Eq. (2.11). Also, we need to have a more definite model of the converter, to firm up expressions for its energy and entropy emission fluxes.

Landsberg considered the case where the converter is a black-body at temperature T_C (Landsberg and Tong 1980). Although this is a more appropriate model for solar thermal conversion, it turns out that, for very high performance, photovoltaic converters emit light similar to a hot body, even although operating at close to ambient temperature. The results are therefore more general than might first appear.

Substituting the corresponding entropy emission expression into Eq. (3.5), the calculated efficiency becomes:

$$\eta = \dot{W}/\dot{E}_S = (1 - \frac{4}{3}T_A/T_S) - T_C{}^4(1 - \frac{4}{3}T_A/T_C)/T_S{}^4 - T_A\dot{S}_G/\dot{E}_S \qquad (3.6)$$

If it is assumed that $\dot{S}_G$ has its minimum value of zero, this expression has its maximum when $T_C = T_A$, giving the Landsberg limit:

$$\eta_L = 1 - \frac{4}{3}T_A/T_S + \frac{1}{3}T_A{}^4/T_S{}^4 \qquad (3.7)$$

This corresponds to an efficiency of 93.3%. While this almost certainly represents an upper bound on solar energy conversion efficiency, it has been claimed that this limit is not attainable even in principle, due to unavoidable entropy production during light absorption (Pawels and De Vos 1981; Marti and Araujo 1996). These claims, however, are based on the assumption of reciprocity between light absorption and emission. As shown elsewhere (Ries 1983), the Landsberg limit can be arbitrarily closely approached, in principle, by time asymmetric systems.

Hence, although the Landsberg limit was derived for quite a specific set of conversion assumptions, specifically a black-body absorber and the assumption of reciprocity between absorption and emission, it represents a far more general limit on solar energy conversion efficiency. Fundamentally, this is because the emission of ambient temperature black-body radiation is the best of all possible options for a converter when zero entropy is produced during the conversion process, a situation fortuitously captured in the space explored by the Landsberg model.

3.5 Black-Body Limit

The Landsberg analysis can be extended to find tighter limits for reciprocal black-body absorbers by taking into account the unavoidable entropy production during the absorption and emission of light by the black-body. As discussed elsewhere (Planck 1959; De Vos and Pawels 1983), the entropy produced in the converter during black-body light absorption is given by:

$$\dot{S}_a = \dot{E}_S(1/T_C - \frac{4}{3}/T_S)$$ (3.8)

The entropy generated within the converter due to its emission of black-body radiation is given by:

$$\dot{S}_e = \frac{1}{3}\dot{E}_C/T_C$$ (3.9)

As mentioned above, the sum of these is only zero if $T_C = T_S$. Inserting Eqs. (3.8) and (3.9) into Eq. (3.6) as a component of $\dot{S}_G$, together with flux $\dot{S}'_G$ representing additional entropy generation beyond this apparently unavoidable component gives:

$$\eta = (1 - T_C^4/T_S^4)(1 - T_A/T_C) - T_A\dot{S}'_G/\dot{E}_S$$ (3.10)

This is quite a reasonable result as can be deduced by considering Fig. 3.3. The first bracketed term represents the nett radiative energy transfer between the sun and the converter (absorber) while the second is the Carnot conversion limit for a source of energy at the converter temperature. The history and properties of this equation have been discussed elsewhere (de Vos 1992). In the absence of any additional entropy generation, maximum efficiency is obtained for the value of T_C that is the solution of the fifth order equation:

$$4(T_C/T_S)^5 - 3(T_A/T_S)(T_C/T_S)^4 - (T_A/T_S) = 0$$ (3.11)

For $T_A/T_S = 0.05$ (300 K / 6000 K), the solution gives T_C/T_S equal to 0.424 or $T_C = 2544$ K, corresponding to a maximum efficiency of 85.4%.

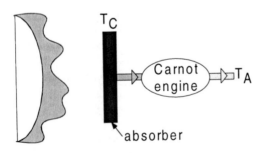

Fig. 3.3: Ideal solar thermal converter with sunlight absorbed by an absorber at temperature T_C, with heat extracted from this absorber converted to electricity by a Carnot converter.

3.6 Multi-Colour Limit

The emission from a black-body at each photon energy, hf, is determined by one parameter, its temperature. Higher efficiency is possible if the emission of light at each energy is optimised separately. Staying with black-body models, a converter which allows this, in principle, is shown in Fig. 3.4. Here, photovoltaics has a real advantage over solar thermal approaches. Using tandem cell stacks, a conceptually equivalent geometry can be implemented much more elegantly, as shown later.

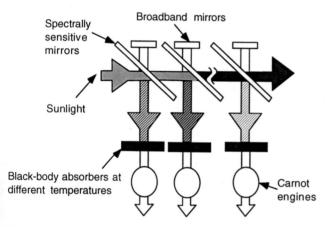

Fig. 3.4: Multicolour converter implemented using black-bodies and mirrors, both spectrally sensitive and broadband. The mirror arrangement shown ensures, in principle, that each black-body absorber both absorbs light over a narrow range of wavelengths, determined by the spectrally sensitive mirrors, and also has a nett emission over the same wavelength range.

The analysis becomes much more complex but, as shown later, results in an optimum temperature for the black-bodies for each photon energy given by:

$$T_{opt} = \frac{T_S}{1+(\frac{kT_S}{hf})\,ln\left\{\frac{1+(1-T_A/T_{opt})(hf/kT_A)-exp(-hf/kT_{opt})}{1+[(1-T_A/T_{opt})(hf/kT_A)-1]\,exp(-hf/kT_{opt})}\right\}}$$

(3.10)

The efficiency is marginally increased to 86.8% from 85.4% for a single black-body, clearly not worth the effort, especially as each black-body absorber in Fig. 3.4 is connected to its own Carnot converter.

We will see later that a photovoltaic cell in combination with a monochromatic filter (Fig. 3.5) can convert energy from a black-body at the Carnot efficiency.

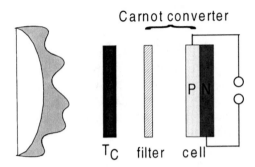

Fig. 3.5: An ideal solar cell combined with an ideal monochromatic light filter acts as an ideal Carnot converter of heat emitted by a black-body.

Hence, each black-body absorber in Fig. 3.4 and its associated Carnot converter can be replaced by an idealised photovoltaic cell of bandgap appropriate for converting its assigned colour (Fig. 5.1 later in the text shows a similar configuration).

An elegant simplification of this modified system is to use the cells to do the filtering as in the tandem stack of Fig. 3.6. This gives the same limiting efficiency as before but with much reduced system complexity.

The author has argued elsewhere that this multi-colour efficiency limit of 86.8% is the highest solar energy conversion efficiency possible in a solar system with reciprocity between light absorption and emission (Green 2001).

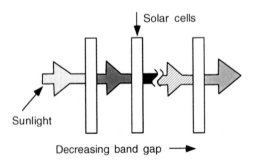

Fig. 3.6: Tandem cell stack giving the same limiting efficiency as the more complex system of Fig. 3.4.

3.7 Non-Reciprocal Systems

Most of the basic laws of physics are time symmetric, in that they stay unchanged if it were possible to reverse the direction of time. This time symmetry is the reason for some of the important reciprocal relationships in optics such as Kirchhoff's Law, which states that the absorptance of a surface equals its exitance (emissivity).

However, it is well known that many of the laws of physics are not reciprocal in the presence of a constant magnetic field. An example is Faraday rotation, where the plane of polarisation of light is changed when it passes through matter in the presence of such a field. Passing this light in the opposite direction does not reverse this rotation, allowing the design of non-reciprocal elements such as circulators, routinely used in microwave circuitry [including the microwave detected photoconductance decay systems often used to measure photovoltaic material quality (Schmidt 2002)].

The basic feature of a circulator is that it can accept radiation from one direction while emitting it in a different, as implied by the circuit symbol in Fig. 3.7(a). Ries has pointed out (Ries 1983) that such circulators can be used to improve solar energy conversion efficiency, in principle, by using the circulators in the configuration of Fig. 3.7(b), bearing some similarity to the earlier Fig. 3.4, although no filtering is involved in the present case.

Rather than being directed back to the sun, the use of a circulator allows the light emitted by the first converter to be passed onto the second, and so on. In principle, each converter then needs to absorb only an infinitesimally small amount of energy, allowing it to be absorbed without entropy generation (Ries 1983).

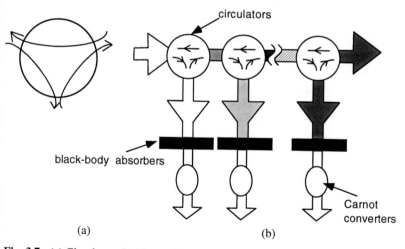

Fig. 3.7: (a) Circulator. (b) Non-reciprocal converter based on circulators.

However, there is no need even to discuss entropy to calculate the limiting efficiency of this combination. Consider one of the black-body converters of Fig. 3.7 at temperature T receiving radiation from the previous converter at temperature $T + \Delta T$. The energy absorbed by the converter is assumed to be converted at the Carnot efficiency $(1 - T_A/T)$ giving a work output:

$$\Delta \dot{W} = \sigma [(T + \Delta T)^4 - T^4](1 - T_A/T) \approx 4\sigma T^3 (1 - T_A/T) \Delta T \qquad (3.11)$$

Integrating all such contributions as $\Delta T \rightarrow 0$ gives:

$$\dot{W} = 4\sigma \int_{T_A}^{T_S} T^3 (1 - T_A/T) dT = 4\sigma [(T_S^4 - T_A^4)/4 - T_A(T_S^3 - T_A^3)/3]$$

$$\qquad (3.12)$$

$$\eta = \dot{W}/\dot{E}_S = 1 - T_A^4/T_S^4 - \frac{4}{3}T_A/T_S + \frac{4}{3}T_A^4/T_S^4$$

$$= 1 - \frac{4}{3}T_A/T_S + \frac{1}{3}T_A^{4}/T_S^4$$

$$\qquad (3.13)$$

or the Landsberg efficiency!

Surprisingly, this derivation requires only the Stefan-Boltzman law and acceptance of the expression for limiting efficiency derived by Carnot without reference to entropy. Importantly, no knowledge of the entropy of radiation is required, so obviously the source of the 4/3 term in the Landsberg derivation.

One issue in this derivation arises from circulator losses. Is it possible in principle to conceive of the lossless circulators assumed in the previous analysis? Ries has suggested a circulator concept based on flicking mirrors (Ries 1983) that allows a more intuitive feel for this aspect than may be possible with Faraday rotation. Further insight is provided by analysing the concept with only a single circulator. This increases the limiting efficiency from 85.4% to 90.0% (Brown and Green 2002), suggesting that circulator efficiencies above 95% will result in performance increase. Since such efficiencies using the mirror arrangements of Ries are at least conceptually feasible, this establishes that non-reciprocal systems can, in principle, exceed the efficiency of those that are reciprocal or time symmetric (Trzeciecki and Hübner 2000).

3.8 Ultimate System

A slight improvement over the previous systems can be obtained with the non-reciprocal photovoltaic system of Fig. 3.8 (Green 2002a). This combines circulators with infinite stacks of tandem cells. Although the algebra is far more complex, the limiting efficiency also reduces to the Landsberg limit of 93.3% in

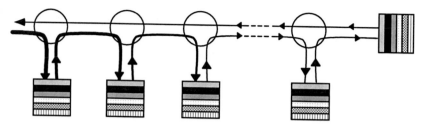

Fig. 3.8: Time asymmetric photovoltaic system using an infinite number of circulators in combination with infinite stacks of tandem cells (each of the stacks shown correspond to solar cells in the configuration of Fig. 3.6).

the limit of an infinite number of circulators. The advantage over Fig. 3.7(b) becomes evident if finite numbers of circulators are involved, with the system of Fig. 3.8 always giving higher limiting efficiency.

The question arises as to whether or not a similar simplification that converted the very complex arrangement in Fig. 3.4 to the much more elegant Fig. 3.6 is possible for the complex structure in Fig. 3.8. If so, this would represent the ultimate photovoltaic converter!

3.9 Omnidirectional Global Converters

All the previous limits apply to converters that convert only the direct component of sunlight, such as in sun-tracking concentrator systems. Most photovoltaic systems, however, are flat plate systems that can convert sunlight reasonably efficiently regardless of its angle of incidence. Although this ability gives enormous practical advantages since these cells can convert diffuse light and sun-tracking also is not required, these advantages come at the expense of reduced limiting efficiency.

The extent of this reduction can be calculated by noting that it does not matter to the efficiency of such an omnidirectional converter whether the light comes highly collimated or spread over all possible angles, with correspondingly higher entropy flux.

For example, if a given total energy flux of 6000K radiation is spread over the whole available illumination hemisphere rather than being confined to the small range of angles subtended by the sun, the associated entropy flux increases by a factor close to 4 (Landsberg and Tonge 1980). The corresponding value of the Landsberg efficiency limit decreases to 73.7% due to a value of the second term in Eq. (3.7) higher by this factor. The other efficiency limits mentioned decrease roughly in the same proportion.

As discussed in Sect. 2.5, it is possible to conceive of schemes that allow the direct component of sunlight to be converted very efficiently while still responding to diffuse light. An example was shown in Fig. 2.7 where concentrator cells are used to convert the direct component, while non-concentrating cells convert the diffuse light, plus that scattered by the optics. The limiting efficiency of such an approach would be a weighted average of the direct and diffuse limits, taking into account the increased entropy per unit energy due to the reduced intensities of each.

3.10 Summary

Various limiting efficiencies for solar conversion have been described. The upper limit on solar energy conversion efficiency is 93.3% obtained for a device that is non-reciprocal in its absorption and emission properties and uses only the direct component of sunlight. The limiting efficiency decreases to 86.8% for systems reciprocal in absorption and emission properties. Both values are well above the limiting value for a standard cell under direct sunlight, which we will show in Chap. 4 is equal to 40.8%.

Table 3.1 compares the different efficiencies mentioned in the text and lists the different conversion options that have been suggested roughly in the order of decreasing limiting efficiency. Many of these options will be discussed in subsequent chapters.

Table 3.1: Limiting efficiencies and possible implementation strategies (after Green 2002b).

η Limit	Direct	Global	Implementation
Landsberg	93.3%	73.7%	circulators antenna?
Multicolour	86.8%	68.2%	hot carrier tandem cells impact ionisation
Black-body	85.4%	53.6%	thermal electric thermophotovoltaics thermionics up- & down-converters
3-level	63.8%	49.3%	3 cell stack impurity PV impurity band up-converters down-converters
2 cells	55.7%	42.9%	2 cell stack
Single junction	40.8%	31.0%	single junctions
Best Laboratory	34%	30%	3 cell tandems
	-	14%	Third Generation

Exercise

3.1 Consider a metal plate with one side painted black so it acts like a nearly perfect absorber of sunlight incident on this surface while the other surface is unpainted and acts as a nearly ideal reflector. Derive algerbraic expressions for:

(a) the nett energy absorbed by the plate as a function of its temperature if sunlight is maximally concentrated onto the plate;

(b) the maximum possible efficiency for converting heat extracted from the plate into electricity;

(c) the maximum solar energy conversion efficiency.

Iteratively, or otherwise, find the conditions giving the maximum possible values for the corresponding efficiency for an effective sun temperature of 6000K and an ambient (sink) temperature of 290K.

3.2 Calculate the limiting efficiency for the non-reciprocal system of Fig. 3.7(b) when only one circulator and two black-body absorbers are used. Assume the sun's temperature is 6000K and the ambient temperature is 290 K.

References

Brown A and Green MA (2002), to be published.

Clausius R (1865), Annalen der Physik 125, 353.

Davies PD and Luque A (1994), Solar thermophotovoltaics: Brief review and a new look, Solar Energy Materials and Solar Cells 33: 11-22.

De Vos A (1992), Endoreversible thermodynamics of solar energy conversion, Oxford University Press, Oxford.

De Vos A and H. Pauwels (1983), Comment on a thermodynamical paradox presented by P. Würfel, J Phys C: Solid State Phys 16: 6897-6909.

Dugdale JS (1996), Entropy and its physical meaning, Taylor and Francis, London.

Green MA (2001), Third generation photovoltaics: Recent theoretical progress, 17th European Photovoltaic Solar Energy Conference and Exhibition, Munich, 14-17.

Green MA (2002a), Third generation photovoltaics: Comparative evaluation of advanced solar conversion options, 29th IEEE Photovoltaic Specialists Conference, New Orleans, May.

Green MA (2002b), Efficiency limits for photovoltaic solar energy conversion, Photovoltaics for Europe Conference and Exhibition on Science, Technology and Application, Rome, October.

Landsberg PT and Tonge G (1980), Thermodynamic energy conversion efficiencies, J Appl Phys 51: R1-R20.

Marti A and Araujo G (1996), Limiting efficiencies for photovoltaic energy conversion in multigap systems, Solar Energy Materials and Solar Cells 43: 203-222.

Pauwels H and De Vos A (1981), Determination of the maximum efficiency solar cell structure, Solid State Electronics 24: 835-843.

Planck M (1959), "The theory of heat radiation", (Dover, New York) [English translation of Planck M (1913), "Vorlesungen über die Theorie der Warmestrahlung", (Leipzig: Barth)].

Ries H (1983), Complete and reversible absorption of radiation, Applied Phys B32: 153.

Trzeciecki M and Hübner W (2000), Time-reversal symmetry in non-linear optics, Phys. Rev. B62: 13888-13891.

Schmela M (2002), Deceptive variety: Market survey on solar modules, Photon International, February, 42-51.

Schmidt J (2002), private communication, June.

Shockley W and Queisser HJ (1961), Detailed balance limit of efficiency of p-n junction solar cells, Journal of Applied Physics 32: 510-519.

Uvarov EB and Chapman DR (1971), A dictionary of science, 4th ed., Penguin Books, Harmondsworth.

4 Single Junction Cells

4.1 Efficiency Losses

Figure 4.1 shows the energy band diagram of a standard p-n junction solar cell including major energy conversion losses when illuminated (Green 1982; Green 1995).

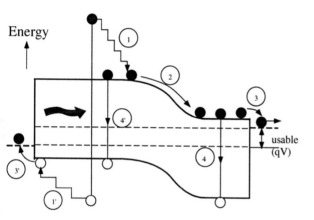

Fig. 4.1: Energy conversion loss processes in a standard single-junction solar cell: (1) lattice thermalisation loss; (2) junction loss; (3) contact loss; (4) recombination loss. A fifth arises from photons that have insufficient energy to be absorbed by the cell.

Basically, the n-type regions of the device (right-hand side of Fig. 4.1) are regions where conduction band electrons can flow easily to and from cell contacts while valence band holes cannot. The p-type regions have opposite properties. This asymmetry causes flows of photogenerated electrons and holes in the opposite directions as indicated. Current flows are described in terms of quasi-Fermi or electrochemical potential gradients (Appendix E). If carriers have high mobility, only small gradients are required to support large current flows.

Assuming infinite carrier mobilities results in zero quasi-Fermi-level gradients as for the broken lines in Fig. 4.1. This assumption greatly simplifies device analysis. A bonus is that this is also consistent with obtaining maximum possible efficiency, since such gradients result in entropy production. As explored elsewhere [(Green 1997) and references therein], there is also a very close relationship between carrier collection efficiency and quasi-Fermi level separation. Any cell designed to collect all photogenerated carriers will automatically satisfy the constant quasi-Fermi separation criterion without the need for the infinite mobility assumption.

Ideally, electrons should flow readily not only in n-type material but also across to the associated metal contact (low contact resistance is desirable!). This corresponds to continuity between the electron quasi-Fermi level and the Fermi level in the metal contact, as indicated to the right of Fig. 4.1. A discontinuity here results in entropy production. Similarly, the hole quasi-Fermi level is ideally continuous with the Fermi level in the metal contact to the p-type region. Combined with the infinite mobility assumption, this forces the quasi-Fermi levels to be constant throughout the device separated by an energy equal to the separation of the contact Fermi levels. This equals qV, where V is the applied voltage.

At the contacts, the above assumptions ensure there is a discontinuity in quasi-Fermi level for minority carriers. By assuming zero minority carrier flow across this contact regardless of the size of this discontinuity (zero surface recombination velocity), entropy production here is also avoided.

These assumptions make the devices very easy to analyse. Since diffusion lengths are also infinite, all photogenerated carriers are collected. It follows:

$$I = q\iint\int^{v} G dV - q\iint\int^{v} U dV \tag{4.1}$$

where I is output current and G and U are the generation and recombination rates per unit volume (Green 1997).

The first term equals the total generation of electron-hole pairs throughout the volume by the incident light and equals I_L, the collected current in the absence of recombination. The second term is usually expressed in terms of carrier recombination lifetimes operating on excess carrier concentrations. Since the Fermi levels are constant throughout the device, it follows that, everywhere in the device, under non-degenerate conditions (Appendix E):

$$np = n_i^2 e^{qV/kT} \tag{4.2}$$

where n and p are electron and hole concentrations at a given point, and n_i is the intrinsic carrier concentration. This expression can be extended to degenerate conditions by defining an effective intrinsic carrier concentration, n_{ie}, with a value chosen to make this relation correct!

In the two limits of low and high injection, U can be very simply expressed in terms of a carrier decay lifetime. In p-type or intrinsic material:

$$U = \frac{n - n_i}{\tau} \tag{4.3}$$

For low injection conditions:

$$\iiint U dV = (e^{qV/kT} - 1)\left[\iint\int \frac{n_{ie}^2}{\tau_n N_A}^P dV + \iint\int \frac{n_{ie}^2}{\tau_p N_D}^N dV \right] \tag{4.4}$$

while, for high injection:

$$\iiint U dV = (e^{qV/2kT} - 1)\iiint \frac{n_{ie}}{\tau} dV \tag{4.5}$$

resulting in the ideal solar cell equation:

$$I = I_L - I_o (e^{qV/NkT} - 1) \tag{4.6}$$

Devices with some regions in low injection and other regions in high injection have contributions to the volume integral of both types, giving an ideality factor, N, that is voltage dependent.

Such a formulation can be used to determine limiting cell performance when non-radiative processes dominate recombination (Green 1984). It is also useful in determining the effect of non-radiative processes upon radiative limits. If the approach is applied to the limiting case where radiative recombination dominates, things seem very straightforward at first since, traditionally (Green 1995):

$$U = B(np - n_i^2) \tag{4.7}$$

B can be deduced from the absorption coefficient using the Shockley-van Roosbroeck equation (van Roosbroeck and Shockley 1954):

$$n_i^2 B = 8\pi c (kT/ch)^3 \int_0^\infty \frac{n^2 \alpha \varepsilon^2 d\varepsilon}{e^\varepsilon - 1} \tag{4.8}$$

where $\varepsilon = hf/kT$, and n and α are the refractive index and absorption coefficient (frequency dependent in general). This allows the same form of equation as in Eq. (4.6) to be derived. In this case, $N = 1$ in both low and high injection!

There are two problems with this formulation. One, discussed later, is that it neglects recombination by stimulated emission that becomes important when qV approaches E_G, the semiconductor bandgap. A second problem is that the formulation ignores the fact that the recombination process itself creates a photon of energy above the bandgap. This photon is capable of being re-absorbed in the semiconductor.

This might be thought to be a minor effect, since such photons will be weakly absorbed compared to the bulk of the more energetic solar photons. However, the photon emitted is emitted in a random direction. This means it automatically gets the full benefits of Lambertian light trapping (Green 1995), regardless of how poorly the cell is designed in this regard for solar photons. Almost all these emitted photons can be re-absorbed or "recycled" in some cases (Exercise 4.1). An elegant approach to solving this problem with the analysis was described in 1961 (Shockley and Queisser 1961).

4.2 Shockley-Queisser Formulation

Shockley and Queisser's approach removed the need to keep track of the recycled photons by switching attention to what was happening outside the cell. They realised that an efficient cell would have to be an efficient absorber of solar photons and hence have properties related to a black-body, at least for energies above the cell's bandgap.

By assuming that an ideal cell did have black-body properties, they were able to very simply calculate the net rate of recombination in the cell taking into account photon recycling effects. By considering such a cell in thermal equilibrium (no external light on it and no voltage applied to it), they realised it would have to be emitting Planckian black-body radiation at these energies, of the same type as treated in Chap. 2.

By attributing all this radiation for energies above the semiconductor's bandgap to band-to-band recombination, they were able to very simply calculate the net rate of recombination events occurring in the cell. This is given by A times the hemispherical black-body emission rate (Eq. (2.9)) for photons of energy higher than the bandgap. A is the area of the cell from which light can be emitted (this is ideally only the front surface area since a reflector can be placed on the rear).

Moreover, From Eqs. (4.2) and (4.7), all radiative recombination in the cell should increase exponentially with applied voltage. Since photon recycling effects would remain proportionately the same, this meant Shockley and Queisser could now calculate the net recombination rate at any voltage. This leads instantly to the ideal solar cell equation with unity ideality factor $(N = 1)$ and I_o given by:

$$I_o = qA\dot{N}(E_G,\infty) \approx qA(\frac{2\pi kT}{h^3 c^2})\left[E_G{}^2 + 2(kT)E_G + 2(kT)^2\right]e^{-E_G/kT} \quad (4.9)$$

The expression on the right can be derived if the -1 term in the denominator of Eq. (2.9) is neglected, a good assumption for $E_G \gg kT$ (the resulting integral is then readily integrated by parts).

The interesting feature of this expression is that I_o no longer depends on cell volume. Although the total number of radiative recombination events increases with volume, photon recycling becomes more effective so that the net number of recombination events stays the same. With the infinite mobility assumption, all photogenerated carriers will be collected. This means no photogenerated carrier will be lost, regardless of how large the cell becomes.

Although no semiconductor has infinite mobility, the parallel multijunction structure of Fig. 4.2 allows the infinite mobility case to be approached arbitrarily

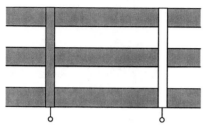

Fig. 4.2: Parallel multijunction solar cell.

closely for materials with finite mobility. The thickness of each multilayer region has to be a lot less than a quantity of the order of $(kTn\mu /J)$ where μ is the mobility of minority carriers of concentration, n, and J is the current density perpendicular to the surface. The lateral extent between contacts has to be less than a similar term for majority carriers, with J the lateral current density in this case.

If the sun is modelled as a black-body at temperature T_s, the current-voltage curve can be expressed as:

$$I = qAf_s\dot{N}(E_G,\infty,T_s)- qA\, f_c\dot{N}(E_G,\infty,T_c)(e^{qV /kT} -1) \qquad (4.10)$$

where f_s and f_c are geometrical factors to be discussed. Equation (4.10) shows that cell properties are described solely in terms of the bandgap in this analysis. For the non-concentrating system analysed by Shockley and Queisser, the solar intercept fraction $f_s = 2.1646 \times 10^{-5}$ while $f_c = 1$, leading to a peak efficiency of 31.0% for $E_G = 1.3$ eV for $T_s = 6000$ K and $T_c = 300$ K.

For a concentrating system, $f_s = 1$. From Eq. (4.10), this means that a larger V is required to reduce I to zero, i.e., a larger V_{oc} is obtained in this case. This increases the peak efficiency to 40.8% for $E_G = 1.1$ eV. The same value is also obtained for the limiting efficiency conversion of direct sunlight at other concentration levels, including non-concentration. In this case, f-factors, f_s and f_c would apply to both sun and cell. The former would be determined by the concentration level and the latter by the angular selectivity of the cell. To collect all direct sunlight, $f_c \geq f_s$. In principle, it is always possible to design a system with $f_c = f_s$, regardless of f-factor value between the extremes mentioned. This means that the same V_{oc} and hence efficiency is obtained regardless of concentration level, for direct light.

The work of Shockley and Queisser represented a major conceptual advance and provided a new set of tools for analysing solar cell performance. One limitation is that it applies only to non-degenerate situations as determined by $(E_G - qV) \gg kT$. This is not a serious limitation for single junction cells but could be for more general photovoltaic converters.

4.3 Hot Photons
(Chemical Potential of Light)

The Bose-Einstein distribution function applies not only to photons, but to other particles (or bound entities) known as bosons. Bosons have integral spin as opposed to fermions that have half-integral spin. Alternatively, bosons can be described by wave functions that are unchanged on interchange of any two particles while those of fermions change sign (Feynman et al. 1965; Bowley and Sanchez 1996).

A more general form of the Bose-Einstein distribution function is:

$$f_{BE} = \frac{1}{e^{(E-\mu)/kT} - 1}$$

(4.11)

where E is the boson energy and μ is its chemical potential. Readers with prior contact with semiconductor devices would be familiar with corresponding expressions for fermions, where the electrochemical potential is closely related to quasi-Fermi levels. The chemical potential is the free energy brought to the system by the addition of one particle which is different from the energy of the particle itself.

For a system at constant temperature, T, and pressure, P, the incremental change in energy in such a system due to variations in particle numbers, N_i, is given by:

$$dE = TdS - PdV + \sum_i \mu_i dN_i$$

(4.12)

Thermal equilibrium corresponds to minimum free energy, so E is invariant to changes in the variables indicated at thermal equilibrium. If photons are the only particles around, as in a cavity, it follows that the chemical potential of photons must be zero since $\mu_{pt} dN_{pt} = dE = 0$.

This property was thought to apply to all light, complicating most thermodynamic treatments prior to about 1980 with some exceptions (Ross 1967). It was then realised that it was reasonable to associate a chemical potential with light emitted by electron-hole recombination (Würfel 1982). This is because at least three particles are involved in the recombination process; the electron, hole and photon. Applying this information to Eq (4.12) near equilibrium, gives:

$$\mu_e dN_e + \mu_h dN_h + \mu_{pt} dN_{pt} = 0$$

(4.13)

For the creation of a single photon by carrier recombination, we have the relationship $dN_e = dN_h = -dN_{pt} = -1$, giving:

$$\mu_{pt} = \mu_e + \mu_h = E_{Fn} - E_{Fp}$$

(4.14)

where E_{Fn} and E_{Fp} are electron and hole quasi-Fermi levels. As already discussed, their difference equals qV for the idealized solar cells being discussed. This gives:

$$f_{BE} = \frac{1}{e^{(E-qV)/kT} - 1}$$

(4.15)

Photons produced by carrier recombination are disturbed from their thermal equilibrium distribution. The light emitted by the cell consists of these "hot" photons. Since photons do not interact with one another, they maintain this hot distribution until they interact with something else. We now need a new theory of black-body emission that takes into account the possibility of non-zero photon chemical potential. We define parameters equivalent to those of Eqs. (2.9) to (2.11) but with this new variable introduced:

$$\dot{N}(E_1, E_2, \mu, T) = \frac{2\pi}{h^3 c^2} \int_{E_1}^{E_2} \frac{E^2 dE}{e^{(E-\mu)/kT} - 1}$$

(4.16)

$$\dot{E}(E_1, E_2, \mu, T) = \frac{2\pi}{h^3 c^2} \int_{E_1}^{E_2} \frac{E^2 dE}{e^{(E-\mu)/kT} - 1}$$

(4.17)

$$\dot{S}(E_1, E_2, \mu, T) = \frac{\dot{E}}{T} - \frac{\mu \dot{N}}{T} - \frac{2\pi k}{h^3 c^2} \int_{E_1}^{E_2} E^2 \ln(1 - e^{-(E-\mu)/kT}) dE$$

(4.18)

Note the extra term in the entropy expression compared to Eq. (2.11). This might have been anticipated from previous discussion (this will tend to reduce the entropy loss due to light emission!). Balancing the number of photons into the cell with the sum of those emitted plus the number of electrons flowing out of the cell, the equivalent of Eq. (4.10) is derived:

$$I = qAf_s \, \dot{N}(E_G, \infty, 0, T_s) + qA(f_c - f_s)\dot{N}(E_{\hat{G}}, \infty, 0, T_c) - qAf_c \dot{N}(E_G, \infty, qV, T_c)$$

(4.19)

The first term on the right is identical to that in the earlier Eq. (4.10) while the third term corresponds to the second. Noting that for $(E - qV) \gg kT$,

$$\frac{1}{e^{(E-qV)/kT} - 1} \approx \frac{1}{e^{(E-qV)/kT}} \approx \frac{e^{qV/kT}}{e^{E/kT}} \approx \frac{e^{qV/kT}}{e^{E/kT} - 1}$$

(4.20)

It follows, under these conditions that:

$$\dot{N}(E_G, \infty, qV, T_c) \approx e^{qV/kT} \dot{N}(E_G, \infty, 0, T_c)$$

(4.21)

This shows the formulations of Eqs. (4.10) and (4.19) are essentially identical under non-degenerate conditions.

A slight difference is in the treatment of the *–1* term in the final brackets of Eq. (4.10). The middle term on the right corresponds to the situation in Fig. 4.3 where, as well as being radiated by the sun, the cell is radiated by black-body radiation from the rest of the hemisphere, assumed to be at the same temperature as the local ambient. This contribution has negligible effect upon most practical calculations but gives the satisfying result that, if the sun were replaced by a segment at the cell's temperature and the cell short-circuited, no current would flow in the cell leads. Equation (4.19) produces exactly the same results as Eq. (4.10), in all but the most extreme cases.

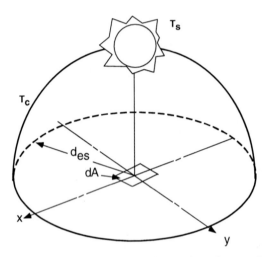

Fig. 4.3: Illustration of radiation environment seen by an element on the cell's surface. As well as the direct sunlight at temperature T_s, the cell seeks sky radiation assumed to be at the ambient temperature T_c.

The normal approach to finding the maximum efficiency, η, of a single junction device is to fix the bandgap, E_G, and find the voltage, V, that maximises the product VI, with I given by Eq. (4.19). The value of E_G is then varied to find η as a function of E_G and hence the overall optimum. This procedure is equivalent to finding $(\delta\eta / \delta V)_{E_G} = 0$ and then the optimum E_G.

However, the derivative $(\delta\eta / \delta E_G)_V$ is much simpler to evaluate than $(\delta\eta / \delta V)_{E_G}$. In fact, for $f_s = f_c$, the former derivative is simply shown to equal zero when $V = E_G (1 - T_c / T_s)/q$, corresponding to a Carnot factor applied to the output voltage (Landsberg 2000). Note that this voltage is not the optimal voltage for fixed E_G but, if the voltage is fixed, maximum power output is obtained for the value of E_G given by the above relationship (Baruch et al. 1992). The two conditions are equivalent only at the optimal value of E_G. Substituting the previous value of V into Eq. (4.19) and finding $(d\eta/dE_G) = 0$ allows the overall optimum η to be found (40.8% when $f_c = f_s$).

Readers interested in silicon cells might wonder how the situation changes when phonons are an integral part of the near band-edge absorption and emission processes. Phonons are also bosons. They are also usually assigned zero chemical potential on the grounds their numbers are not conserved. However, "hot" phonons are also possible, corresponding to finite chemical potential, μ_r (Würfel et al. 1995; Würfel 1995).

Going through the same process as before, the chemical potential of an emitted photon would equal $\mu_e + \mu_h \pm \mu_r$ depending on whether a phonon was absorbed or emitted. However, since, unlike photons, phonons can interact with one another and split to give multiple phonons and so on, it is difficult to disturb phonon populations from equilibrium. Hence taking μ_r equal to zero is valid in all but extreme cases. This means that exactly the same theory applies when phonons are involved (Würfel et al. 1995).

Although the chemical potential formulation is a conceptual advance over the Shockley-Queisser approach, it also is not valid under degenerate conditions *(qV > E_G)*. Under these situations, Eqs (4.11) and (4.15) predict infinite occupation probabilities for some energy and negative values for others. Both would seem unphysical.

Moreover, both formulations assume an abrupt transition at the band-edge. Photons of energy above the bandgap are absorbed, while those below it are not. Sub-bandgap absorption processes involving the assistance of phonons would also occur in direct bandgap material, creating additional electron-hole pairs. Is this a help or a hindrance?

Also, it is clear that processes involving more than one phonon are likely to be occurring, although very weak. Under the above analysis, if many phonons were involved, the photons emitted in such cases would have a chemical potential equal to or larger than their energy, with the unphysical consequences previously noted. Clearly a more complete formulation is possible.

4.4 Einstein Coefficients

While investigating the equilibrium between black-body radiation and transition rates between allowed energy levels in atoms, Einstein, in 1917, discovered an unusual but unavoidable consequence of the Planckian formula (Einstein 1917).

Einstein considered the rate of transition between any two discrete atomic energy levels (Fig. 4.4)

Suppose a collection of N_n atoms are in the state represented by an electron in level n and N_m are in the state represented by an electron in level m. Einstein knew that if these atoms were in equilibrium with the thermal radiation, the relative number in such states would be given by:

$$\frac{N_m}{N_n} = e^{-\Delta E_{mn}/kT}$$

(4.22)

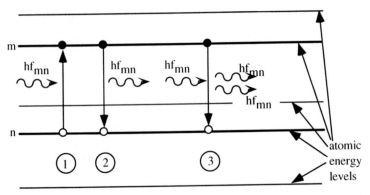

Fig. 4.4: Transitions between atomic levels considered by Einstein in his 1917 paper.

He now considered how these atoms would interact with the black-body radiation with which they were in equilibrium. Photons of the correct frequency ($hf_{mn} = \Delta E_{mn}$) would be converting atoms in state n to those in state m at a rate given by:

$$R_{n \to m} = N_n B_{nm} u_V (hf_{mn}) \tag{4.23}$$

where B_{nm} is a proportionality constant describing the ease of this excitation process and u_V is the black-body energy density given by Eq. (2.6). Einstein then considered the rate that atoms in state m would convert to those in state n. One component would just be the spontaneous emission of radiation as in process 2 of Fig. 4.4 given by $N_m A_{mn}$, where A_{mn} is another proportionality constant. To get the correct result from his calculation, he had to assume a second light emission process dependent on light intensity (process 3 of Fig. 4.4) to which he assigned a rate $N_m B_{mn} u_V (hf_{mn})$ giving:

$$R_{m \to n} = N_m [A_{mn} + B_{mn} u_V (hf_{mn})] \tag{4.24}$$

Equating these rates and applying Eq. (4.23) gives:

$$u_V (hf_{mn}) = \frac{A_{mn} / B_{nm}}{e^{hf_{mn} / kT} - B_{mn} / B_{nm}} \tag{4.25}$$

This could only agree with Planck's formula if, importantly, B_{mn} was finite and equal to B_{nm}. Moreover:

$$A_{mn} / B_{nm} = 8\pi hf_{mn}^3 / c^3 \tag{4.26}$$

This was an enormous amount of information to deduce from such a straightforward calculation!

Note that the -1 term in the Bose Einstein distribution function arises from the stimulated emission (process 3). Note also that the ratio of stimulated to spontaneous emission is identical to the occupation probability, f_{BE}. The rate of absorption by process 1 is $exp(hf/kT)$ times higher $[f_{BE}/(1 - f_{BE})]$ than the rate of stimulated emission.

Two other points are worthy of note. The first is that, while a photon emitted by spontaneous radiation is emitted in a random direction, that emitted by stimulated emission is emitted into the same "mode" as the stimulating photon. This means that it is a carbon copy in all respects, being identical in frequency, polarisation, direction and phase. The second point is that the relative strengths of the processes, when expressed in terms of f_{BE}, retain the same form when a chemical potential can be associated with the light.

This gives rise to prospects for laser action. If the chemical potential were to become higher than the photon energy, stimulated emission would become stronger than normal absorption, amplifying light in its path through the material involved.

4.5 Photon Boltzmann Equation

The Boltzmann equation is a classical equation for particle transport derived by Ludwig Boltzmann in 1872. It describes how a distribution function, f (similar to f_{BE}), might be expected to evolve in space and time. It can be derived quite simply by again considering a 6-dimensional space $xyzv_xv_yv_z$, where v is particle velocity (Bailyn 1994).

A point r_1, v_1, in this space and a small volume $dx_1\, dy_1\, dz_1\, dv_{1x}\, dv_{1y}\, dv_{1z}$ about it are considered. Neglecting collisions, the particles at the point $r_1 - v_1\, dt$, $v_1 - F_1\, dt/m$ at time $t - dt$ would have moved to r_1, v_1 in time dt. They would be the sole occupants of the small volume mentioned at time t since the particles originally there would have moved out (since they have velocity v_1). As a first approximation:

$$f(r_1, v_1, t) = f[(r_1 - v_1\, dt), (v_1 - F_1\, dt/m),(t - dt)]$$
$$= f(r_1, v_1, t) - v_1\, dt\,.\,(\delta f_1/\delta r) - (F_1/m)dt\,.\,(\delta f_1/\delta v) - (\delta f_1/\delta t)\, dt + ... \qquad (4.27)$$

where $f_1 \equiv f_1 (r_1, v_1, t)$. Note that at any position r_1, f_1 may have a different value for different values and directions of particle velocity, a point that will be further explored in Sect. 4.8.

If nothing else happened, the three terms on the right would have to add to zero. However, additional particles enter and leave the small volume due to scattering by collisions. This gives the Boltzmann equation in its standard form:

$$\delta f_1/\delta t + v\,.\,(\delta f_1/\delta r)+(F_1/m)\,.\,(\delta f_1/\delta v)=(\delta f_1/\delta t)_{coll} \qquad (4.28)$$

The Boltzmann equation does have limitations that are summarised elsewhere (Bailyn 1994). Customising for photons, photon velocity does not change due to forces but can change due to refractive index changes. Neglecting this effect, the customised version in the steady state is:

$$\upsilon_G \cdot (\delta f_{pt}/\delta r) = \upsilon_G \cdot \nabla f_{pt} = (\delta f_{pt}/\delta t)_{coll} \tag{4.29}$$

where υ_G is the group velocity of the photon, the velocity of photon energy propagation, equal to c/n in isotropic material. Absorption and emission processes are the main collision processes changing the number of photons in the volume of the 6-dimensional space involved.

Considering the case of photons in a semiconductor, Fig. 4.5 shows two representative series of processes that might be involved. One type involves interband excitation to create electron-hole pairs and the other involves intraband transitions, each with their associated spontaneous and stimulated inverse processes. Let $P_{Ci,Vj}$ be the coefficient describing the rate of exciting an electron state Vj in the valence band when fully occupied to a completely empty state Ci in the conduction band. By the same type of argument as for the Einstein coefficients, it follows that $P_{Ci,Vj} = P_{Vj,Ci}$ while the corresponding coefficient for stimulated emission is $f_{pt} P_{Vj,Ci}$. Similar coefficients $P_{Bi,Bj}$ can be defined for the intraband processes. This gives:

$$(\delta f_{pt}/\delta t)_{coll} = \sum_{i,j} [\, P_{Vj,Ci}(\,f_{pt}+1)f_{Ci}(1-f_{Vj}) - P_{Ci,Vj}f_{pt}f_{Vj}(1-f_{Ci})\,]$$
$$+ \sum_{i,j} [\, P_{Bj,Bi}(\,f_{pt}+1)f_{Bi}(1-f_{Bj}) - P_{Bi,Bj}f_{pt}f_{Bj}(1-f_{Bi})\,] \tag{4.30}$$

where f_{Ci}, f_{Vj} and $f_{Bi,j}$ are the probabilities of finding the respective states occupied by electrons as given by the Fermi-Dirac distribution function, e.g.:

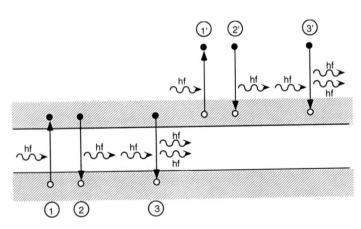

Fig. 4.5: Possible absorption and emission process for a photon of energy above the bandgap.

$$f_{Ci} = \frac{1}{e^{(E_{Ci}-\mu_c)/kT}+1} \tag{4.31}$$

where μ_c is the electrochemical potential of electrons in the conduction band. Note that:

$$(1-f_{Ci}) = f_{Ci}e^{(E_{Ci}-\mu_C)/kT} \tag{4.32}$$

The negative terms in Eq. (4.30) can be identified as arising from band-to-band and free carrier absorption. Combining with the contribution to the positive term from stimulated emission and noting $P_{Ci,Vj} = P_{Vj,Ci}$, this allows net absorption coefficients α_{CV} and α_{FC} to be defined as:

$$\alpha_{CV} = \upsilon_G \sum_{i,j} P_{Vj,Ci}\,(f_{Vj} - f_{Ci}) \tag{4.33}$$

$$\alpha_{FC} = \upsilon_G \sum_{i,j} P_{Bj,Bi}\,(f_{Bj} - f_{Bi}) \tag{4.34}$$

At low excitation levels where stimulated emission is not important, the former absorption coefficient is usually regarded as a material constant since $f_{Vj} \approx 1$ and $f_{Ci} \approx 0$. The free carrier coefficient depends on carrier concentration through the higher of the occupancy functions f_{Bj}. Both coefficients are usually positive. However, the former can become negative under "population inversion", such as desired for lasing. Re-writing Eq. (4.30) in terms of these parameters gives, after manipulation, noting Eq. (4.32) and the equivalent for f_{Vj}:

$$\mathbf{u}.\nabla f_{pt} = \alpha_{CV}\,[\,f_{BE}\,(\mu_{CV}) - f_{pt}\,] + \alpha_{FC}\,[\,f_{BE}\,(0) - f_{pt}\,] \tag{4.35}$$

where $\mathbf{u}$ is the direction of light propagation and $f_{BE}\,(\mu)$ is the Bose-Einstein distribution function for a chemical potential μ as in Eq. (4.11). When $\nabla f_{pt} = 0$, and $\mu_{CV} = 0$, this has the equilibrium solution:

$$f_{pt}^{eq} = f_{BE}(0) = \frac{1}{e^{hf/kT}-1} \tag{4.36}$$

i.e., the Bose-Einstein distribution (as it must since the above essentially involves working through Einstein's analysis from the other direction). For $\mu_{CV} \neq 0$ but $\alpha_{FC} = 0$, the solution is:

$$f_{pt}^{eq} = f_{BE}(\mu_{CV}) = \frac{1}{e^{(hf-\mu_{CV})/kT}-1} \tag{4.37}$$

i.e., the expression derived when a chemical potential can be associated with the photons. In the general case:

$$f_{pt}^{eq} = \frac{\alpha_{CV}f_{BE}(\mu_{CV})+\alpha_{FC}f_{BE}(0)}{\alpha_{CV}+\alpha_{FC}} \tag{4.38}$$

This provides the basis for investigating the energy threshold assumption used in the previous analysis. At the threshold, α_{CV} changes from being small to being large, compared to other absorption processes. Light emitted at energies below the bandgap would retain its normal thermal characteristics.

Two treatments of the sub-bandgap properties of the cell are possible. If α_{FC} were treated as finite, it would be possible for the cell to act as an ideal thermal black-body with zero voltage applied to it. If $\alpha_{FC} = 0$, the cell would act as a spectrally sensitive black-body, being transparent to light of sub-bandgap energy.

If illuminated by black-body light at the same temperature as the cell, there would be little difference between the two cases. The same sub-bandgap light would either be emitted or reflected. For illumination from higher temperature sources, less would be absorbed in the second case and the cell would be heated less.

Combining Eqs. (4.35) and (4.38) gives an alternative version of the former:

$$\mathbf{u}.\nabla f_{pt} = (\alpha_{CV} + \alpha_{FC})(f_{pt}^{eq} - f_{pt}) \tag{4.39}$$

Equation (4.30) implicitly assumes that only photons are contributing to the transitions involved. Surprisingly, the same Eqs. (4.35) and (4.39) is derived even if other particles such as phonons are also involved. The ratio of stimulated emission to absorption for the band to band processes in Eq. (4.30) is given by:

$$R = \frac{f_{Ci}}{(1 - f_{Ci})} \cdot \frac{(1 - f_{Vi})}{f_{Vi}} = e^{(-hf + \mu_{CV})/kT} \tag{4.40}$$

due to Eq.(4.32). The corresponding ratio is given by, in a far more general case (Landsberg 1967), where multiple states, g in total, gain an electron and multiple states, l in total, lose an electron and multiple phonon absorption processes, a, and emission processes, e, are involved:

$$R = \prod_a \frac{f_a}{1 + f_a} \prod_e \frac{1 + f_e}{f_e} \prod_g \frac{f_g}{1 - f_g} \prod_l \frac{1 - f_l}{f_l} \tag{4.41}$$

Since bosons satisfy a similar relationship to Eq. (4.32), this reduces to:

$$R = exp\{[(\sum_e (\varepsilon_e - \mu_{\Gamma e}) - \sum_a (\varepsilon_a - \mu_{\Gamma a}) + \sum_g (E_g - \mu_g) - \sum_l (E_l - \mu_l)]/kT\} \tag{4.42}$$

where ε are the phonon energies and μ are chemical potentials. Energy balance gives:

$$hf + \sum_g E_g + \sum_e \varepsilon_e = \sum_l E_l + \sum_a \varepsilon_a \tag{4.43}$$

Substituting back into Eq. (4.41) gives:

$$R = exp[(\sum_l \mu_l - \sum_g \mu_g + \sum_a \mu_a - \sum_e \mu_e - hf)/kT] \qquad (4.44)$$

If phonons are assumed to have zero chemical potential and the loss of one electron from the conduction band to the valence band is being considered, this reduces to the final value of Eq. (4.40).

4.6 General Cell Analysis

Using the results of the previous section, cell analysis can be generalised to model the transition between free carrier and band-to-band absorption more realistically. Phonon participation allows band-to-band absorption to be finite even for photon energies below the bandgap.

For thick devices, the equilibrium photon occupancy given by Eq. (4.38) gives the total photon emission by the device. For energies where α_{CV} is much larger than α_{FC}, this essentially equals f_{BE} (μ_{CV}) as for the simpler analysis of Sect. 4.3. Below the bandgap where α_{CV} becomes rapidly smaller than α_{FC}, the equilibrium photon occupancy decreases from this value towards the thermal value f_{BE} (0).

Equation (4.39) is very similar to the simpler Beer's law for photon absorption (Green 1982). Only a fraction $\alpha_{CV}/(\alpha_{CV} + \alpha_{FC})$ of absorbed photons, however, will result in electron-hole pair generation, with the remainder giving their energy to free carriers.

Also, photon populations do not decay to zero but to f_{pt}^{eq} . This occupancy function will therefore determine the photon state occupancy of the emitted light in a thick device, as discussed in more detail in a later section. This emission is made up of two terms corresponding to the two terms in Eq. (4.38). The first corresponds to emission by recombination of electrons between bands while the second is due to de-excitation within the same band. Electrons need to be supplied only to replenish the former emission.

In the presence of free carrier absorption, we do have to pay more attention to what is happening inside the device. Section 4.1 described how the total radiative recombination rate in a device could be calculated using Eqs. (4.7) and (4.8). In the Shockley-Queisser case we did not have to worry about these, just the smaller number that were actually emitted. The difference between these are recycled in an ideal device.

However, when free carrier absorption is important, a fraction $\alpha_{FC}/(\alpha_{CV} + \alpha_{FC})$ of these normally recycled photons will be absorbed by free carriers and additional current will be required to account for these. This will be reduced slightly by thermally emitted photons that are absorbed by creating electron-hole pairs. The cell current therefore required to support its light emission is given by:

$$I' = \frac{2\pi q A}{h^3 c^2} f_c \int_0^\infty [\alpha_{CV} / (\alpha_{CV} + \alpha_{FC})] f_{BE}(qV) E^2 dE$$

$$+ \frac{8\pi q}{h^3 c^2} \int_0^{Vol} \int_0^\infty n^2 [\alpha_{CV} \alpha_{FC} / (\alpha_{CV} + \alpha_{FC})][f_{BE}(qV) - f_{BE}(O)] E^2 dE dVol$$

$$- \frac{2\pi q A}{h^3 c^2} f_c \int_0^\infty [\alpha_{CV} \alpha_{FC} / (\alpha_{CV} + \alpha_{FC})^2][f_{BE}(qV) - f_{BE}(O)] E^2 dE$$

$$(4.45)$$

The first term represents the current to support the emitted photon flux, the second represents the current required to support the imperfect recycling of internally generated photons in a cell of refractive index n, calculated by a minor generalization of the Shockley-van Roosbroeck equation (van Roosbroeck and Shockley 1954), while the third term represents a minor correction to the second to account for the fact that some of the internally generated photons do get emitted. The final more general form of Eq. (4.19) for the solar cell current-voltage characteristics is:

$$I = \frac{2\pi q A}{h^3 c^2} \{ f_s \int_0^\infty \frac{[\alpha_{CV} / (\alpha_{CV} + \alpha_{FC})] E^2 dE}{e^{E/kT_s} - 1} + (f_c - f_s) \int_0^\infty \frac{[\alpha_{CV} / (\alpha_{CV} + \alpha_{FC})] E^2 dE}{e^{E/kT_C} - 1}$$

$$- f_c \int_0^\infty [\alpha_{CV} / (\alpha_{CV} + \alpha_{FC})] f_{pt}^{eq} E^2 dE$$

$$- \int_0^{Vol} \int_0^\infty 4n^2 [\alpha_{CV} \alpha_{FC} / (\alpha_{CV} + \alpha_{FC})][f_{BE}(qV) - f_{BE}(0)] E^2 dE dVol / A \}$$

$$(4.46)$$

4.7 Lasing Conditions

Under general conditions, stimulated emission exceeds the corresponding absorption process in strength if $\mu_{CV} > hf$. This means photon density would build up with distance through the semiconductor, if parasitic absorption losses are small, in accordance with Eq. (4.35). What would happen within a solar cell under such conditions? Photon populations at energies where $hf < \mu_{CV}$ would increase enormously, concentrating most of the emitted light at such wavelengths. The very sensitive dependence on μ_{CV} would curtail the open circuit voltage to values only marginally above E_G/q, in general.

For a thick device, there would be one critical wavelength for which the expression of Eq. (4.38) first approached infinity. The emission of light from the cell that balanced the incoming solar radiation would be concentrated at this and surrounding wavelengths. In the case where $\alpha_{FC} = 0$ (i.e., no optical losses) and $\alpha_{CV} = 0$ below the bandgap, μ_{CV} would be limited to a value just below E_G/q, with photon energy concentrated at the latter value.

For cells of finite thickness, there would be no requirement for Eq. (4.35), or, equivalently, Eq. (4.39), to reach an equilibrium value. Values of μ_{CV} further above E_G/q could be reached. A very simple finite geometry has been analysed elsewhere (Parrott 1986). The conclusion from this analysis is that the open-circuit voltage is likely to be affected by these considerations for such thin devices, but not the maximum power point voltage.

Another interesting point concerns sub-bandgap photons. The condition for stimulated emission exceeding the corresponding absorption process is met, even for solar cells of quite modest performance. Why don't we see something unusual at sub-bandgap wavelengths for these devices?

The first term on the denominator of Eq. (4.38) will be negative under such conditions. There would seem to be no fundamental reason why the critical condition mentioned previously could not occur first for such sub-bandgap photons. The energy emitted by the cell would then occur at far infrared wavelengths!

The reason this does not seem to occur is that the achievable negative magnitude of α_{CV} decreases very rapidly with decreasing hf. For silicon, the involvement of each additional optical phonon of 64 meV energy decreases absorption by about 180 times (Green 1995). Hence relatively close to the band edge, the absorption coefficient ratio is increasing from very small values approximately as the exponential of $2hf/kT$. From Eq. (4.33), this transfers to the achievable negative magnitude of α_{CV}. Hence going to lower photon energies decreases the probability of lasing, even though stimulated emission is occurring!

4.8 Photon Spatial Distributions

One feature, not yet fully explored, deals with the directional aspects of light. Thermal radiation is isotropic, but direct sunlight or laser light can follow a highly directional path through the device being analysed. As previously noted in Sect. 4.5, particle occupancy functions can depend on the value and direction of particle velocity. In general (Van Roos 1983), the distribution function for photons f_{pt} (r, u, t, hf) can be defined as a measure of the number of photons of energy hf at position r at time t moving in a direction u to within a solid angle $d\Omega$.

For the situations examined in the previous section, this distinction has not been important. This is not generally the case in solar cells or other photo-electronic devices such as semiconductor lasers. Two situations often encountered in solar cell work require different treatment (Fig. 4.6).

The first case of a planar device appears simple to analyse, but can be quite complex in reality. This is due to the face that, due to Snell's law, light from the illuminating source is quite directional in the semiconductor in this case, as opposed to the recycled light. For the small geometries likely to be of interest for third generation solar cells, interference effects are also likely to be important (Sheng 1984).

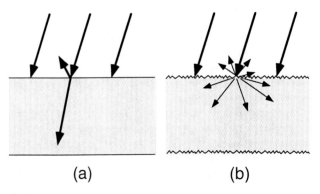

Fig. 4.6: Light incident on (a) a planar surface and (b) a diffusely scattering (Lambertian) surface.

The second case of a diffusely scattering surface is not only simpler to analyse if ideal Lambertian properties are assumed (Green 2002), it also is likely to produce better solar cells.

To get a feel for the differences, the photon distributions in thick cells of both types will be compared. This comparison will be for the case of Fig. 4.3 where the sun is directly overhead with voltage V across the cell. Well in the bulk of the device, the photon occupancy will reach an equilibrium value given by Eq. (4.38).

In the case of the planar device, taking the z direction perpendicular to the cell surface, x and y derivatives are zero by symmetry. Defining an excess directional component of the photon distribution function:

$$f'_{pt} = f_{pt} - f_{pt}^{eq} \qquad (4.47)$$

Equation (4.39) reduces to:

$$u \frac{df'_{pt}}{dz} = -(\alpha_{VC} + \alpha_{FC})f'_{pt} \qquad (4.48)$$

where $u = cos\theta$, where θ is the angle to the normal in the cell. This has the solution:

$$f_{pt} = f_{pt}^{eq} + f'_{pt0_0} e^{-(\alpha_{VC} + \alpha_{FC})z/cos\theta} \qquad (4.49)$$

where f'_{pt0} is the excess function at $z = 0$ and α_{VC} and α_{FC} are net absorption coefficients that can become zero or negative, once stimulated emission becomes important. Figure 4.7 shows typical distributions for a thick device for different values of θ.

Note that directional distributions are continuous across the cell interface for directions where photons are not reflected and such transmission is possible i.e., photons moving within the escape cone. A fraction equal to $1/n^2$ times the number incident isotropically on the interface lie in this cone and will escape, assuming zero reflectance within the escape cone.

The situation for the cell with diffuse surfaces is somewhat different. Regardless of the angle of incidence of the light for an ideal Lambertian surface, its energy is scattered isotropically upon entering the cell. The directional photon distribution function outside the cell, constrained to the solid angle subtended by the sun in the non-concentrating case, is scattered into the hemisphere. Moreover, for light striking the cell surface from the other direction ($\theta > 90°$), only a fraction $1/n^2$ will escape, at least at the macroscopic level (Yablonovitch and Cody 1982; Yablonovitch 1982).

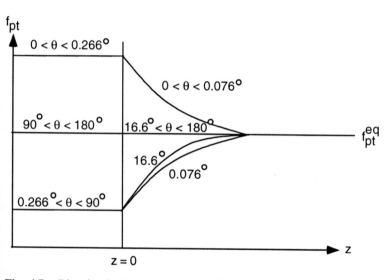

Fig. 4.7: Directional photon state occupation probability as a function of the distance through the cell for different directions specified in terms of θ, the angle to the cell normal (schematic only assuming zero surface reflection) ($n = 3.5$).

Within the cell, the distribution function depends on the direction relative to the normal to the cell surface (Green 2002). The effective absorption coefficient is, on average, multiplied by 2 due to the oblique passage of the average light beam. Figure 4.8 schematically shows these features.

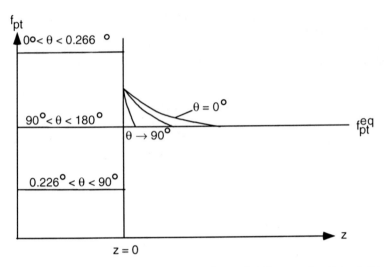

Fig. 4.8: Directional photon state occupation probability as a function of distance through the cell for diffuse (Lambertian) cell surfaces. θ is the angle to the normal to the cell surface.

4.9 Effect of Sample Thickness

Absorption coefficients as defined are independent of device voltage for low values of this voltage, but decrease in magnitude and can even become negative in sign when populations are inverted (higher occupancy in high energy states than in lower energy states).

Equation (4.39) shows that alone any optical path, the photon occupancy decays to the equilibrium value in a way that will be independent of voltage bias, if this is small. For thick devices, the equilibrium value will be reached and determine light emission, as in Figs. 4.7 and 4.8. For thin devices, by the time the rear of the device is reached, the decay may not have reached equilibrium. The optimum design is then to have an ideal rear reflector with negligible loss, which will allow the photon occupancy to be maintained for photons reflected in the opposite direction.

The decay towards the equilibrium value will continue in this direction. By the time the light in this direction reaches the front surface, it still may not have reached equilibrium. The higher value corresponds to increased light output compared to the case of a thick device, due to incomplete absorption of the incoming light.

4.10 Thermodynamics of Single Junction Cell

Considering the system in Fig. 3.2, the energy and entropy balances are, as before:

$$\dot{E}_S = \dot{E}_C + \dot{W} + \dot{Q} \tag{4.50}$$

$$\dot{S}_S = \dot{S}_G + \dot{S}_C + \dot{Q}/T_A \tag{4.51}$$

Combining gives:

$$\dot{W} = \dot{E}_S - \dot{E}_C - T_A (\dot{S}_S - \dot{S}_C + \dot{S}_G) \tag{4.52}$$

In this case, the above terms will be used to refer to quantities integrated from E_G to infinity, multiplied by the appropriate f-factor, e.g., assuming the cell is at ambient temperature:

$$\dot{E}_C = f_c \dot{E}(E_G, \infty, qV, T_A) \tag{4.53}$$

Again, the entropy generated when the cell emits and absorbs light needs to be considered. Figure 4.9 shows the corresponding entropy balances.

The situation is slightly altered from the case of thermal black bodies in that the entropy associated with the particle flows have to be taken into account due to the finite chemical potential $(\mu = qV)$ associated with the photons interacting with electrons and holes in the semiconductor. Note:

$$\dot{S}_{Ge} = \dot{S}_C - (\dot{E}_C - \mu\dot{N}_C)/T_C \tag{4.54}$$

$$\dot{S}_{Ga} = (\dot{E}_S - \mu\dot{N}_S)/T_C - \dot{S}_S \tag{4.55}$$

Inserting into Eq. (4.50) with $T_C = T_A$ gives:

$$\dot{W} = \dot{E}_S - \dot{E}_C - (\dot{E}_S - \mu\dot{N}_S - \dot{E}_C + \mu\dot{N}_C) - T_A\dot{S}_G' \tag{4.56}$$

$$= \mu(\dot{N}_S - \dot{N}_C) - T_A\dot{S}_G' \tag{4.57}$$

In the absence of other entropy generation processes, this corresponds to the expression given in Sect. 4.3. An examination of the terms involved shows that one of the major loss process is the entropy generation on absorption as given by Eq. (4.55).

Entropy is also generated due to finite carrier mobilities (Parrott 1992; Parrott 1982):

$$\dot{S}_{Gm} = \int [(J_e^2 /\sigma_e + J_h^2 /\sigma_h)/T_c]dV \tag{4.58}$$

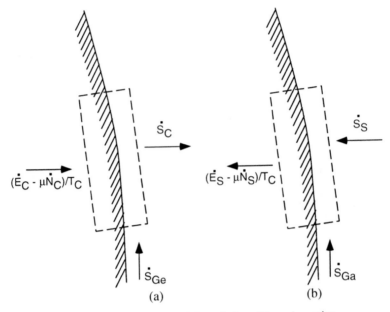

Fig. 4.9: (a) Entropy production on light emission; (b) on absorption.

where J_e and J_h are the carrier current densities at any point within the cell volume, V, and σ_e and σ_h are the carrier conductivities (e.g., $\sigma_e = q\mu_e n$). Finite non-radiative recombination rates also result in a volume dependent entropy production term. Even though the ideal efficiency does not depend on volume, such volume dependent parasitic terms emphasise the need to keep the design compact.

The contacts can provide another source of entropy generation given by (Parrott 1982):

$$\dot{S}_{Gc} = \frac{1}{qT_c}\int(\,\delta E_{Fn}J_e + \delta E_{Fp}J_p\,)\bullet dA \qquad (4.59)$$

where δE_{Fn} and δE_{Fp} are quasi-Fermi energy discontinuities across the contact to semiconductor interfacial area A. For majority carriers such a discontinuity would represent contact resistance. For minority carriers, such discontinuities are related to contact recombination velocity. For zero recombination velocity, the discontinuity is large but the corresponding minority carrier current flow is zero, giving zero entropy production.

Usually, the quasi-Fermi energy discontinuity creates an imbalance in carrier distributions on either side of the interface that drives current flow across it. In some situations, when energy is absorbed, such as for photoexcitation across an interface, current can flow against a quasi-Fermi energy discontinuity.

Exercises

4.1 Derive an algerbraic expression for the fraction of photons recycled in a solar cell as a function of its thickness. Assume a constant refractive index and an absorption coefficient that varies rapidly from zero for photon energies below the cell bandgap to a constant value α_0 for solar photon energies above it. Calculate numerical values for the case of a bandgap of 1.4 eV, refractive index of 3.5, and a constant absorption coefficient of 2 x 10^4/cm, at 300 K.

4.2 For the case where $f_C = f_S$, write down Eq. (4.19) and the corresponding expression for power output, P, in their integral forms. Find the value of V that gives a zero value for $(\delta P/\delta E_G)_V$. Explain what this value means.

4.3 (a) Using one of the series expansions of the Bose-Einstein integrals given in Appendix C, calculate the maximum short-circuit current for a solar cell of the same bandgap as silicon (1.124 eV at 300 K), assuming zero absorption for photon energy below the bandgap energy for: (i) a non-concentrating cell and also; (ii) one at the maximal concentration level. Assume the sun's radiation can be modelled as that from a black-body at 6000 K and that the cell temperature is 300 K.
(b) Calculate the open-circuit voltages in the radiative limit for both the above two conditions using both the Shockley-Queisser and chemical potential formulations of Sects. 4.2 and 4.3 respectively. Which approach would predict the higher cell fill factor?
(c) What is the maximum open-circuit voltage predicted by both approaches at a concentration level of 300 suns?

References

Bailyn M (1994), A Survey of Thermodynamics, AIP Press, New York.

Baruch P, Landsberg PT and de Vos A (1992), Thermodynamic limits to solar cell efficiencies for various illumination conditions, Conf. Record, 11th E.C. Photovoltaic Solar Energy Conference, Montreux, October, 283-286.

Bowley R and Sanchez M (1996), Introductory Statistical Mechanics, Oxford University Press, Oxford.

Einstein A (1917), Zur Quantentheorie der Strahlung, Physik. Z 18:121 [English translation in D.L. Van der Waerden (ed.) (1967), Sources of Quantum Mechanics, Dover, New York, 63-77 and in Ter Haar D (1967), The Old Quantum Theory, Pergamon, Oxford, 167-183].

Feynman RP, Leighton RB and Sands M (1965), The Feynman lectures on physics, Addison-Wesley, Reading MA, Vol. III, Sect. 4.1.

Green MA (1982), Solar Cells: Operating Principles, Technology and System Applications, Prentice-Hall, New Jersey. (Reprinted 1986, 1992) (available from the author).

Green MA (1984), Limits on the open circuit voltage and efficiency of silicon solar cells Imposed by Intrinsic Auger Processes, IEEE Trans Electron Devices ED-31: 671-678.

Green MA (1995), Silicon Solar Cells: Advanced Principles and Practice, (Bridge Printery, Sydney) (available from the author).

Green MA (1997), Generalized relationship between dark carrier distribution and photocarrier collection in solar cells, J App Phys 81: 268-271.

Green MA (2002), Lambertian light trapping in textured solar cells and light emitting diodes: analytical solutions", Progress in Photovoltaics, 10: 235-241.

Landsberg PT (1967), The condition for negative absorption in semiconductors, Phys Stat Sol 19: 777.

Landsberg P (2000), Efficiencies of solar cells: where is Carnot hiding?, Conf. Record, 16th European Photovoltaic Solar Energy Conference, Glasgow, May.

Parrott JE (1982), Transport theory of semiconductor energy conversion, J Appl Physics 32: 9105.

Parrott JE (1986), Self-consistent detailed balance treatment of the solar cell, IEE Proc 133 : 314-318.

Parrott JE (1992), Thermodynamics of solar cell efficiency, Solar Energy Materials and Solar Cells 25: 73-85.

Ross RT (1967), Some thermodynamics of photochemical systems, J Chem. Phys 46: 4950-4593.

Sheng P (1984), Optical absorption of a thin film on a Lambertian reflector substrate, IEEE Trans on Electron Devices ED-31: 634-636.

Shockley W and Queisser HJ (1961), Detailed balance limit of efficiency of p-n junction solar cells, J App Phys 32: 510-519.

van Roosbroeck W and Shockley W (1954), Photon-radiative recombination of electrons and holes in germanium, Phys Rev 94: 1558-1560.

von Roos O (1983), Influence of radiative recombination on the minority-carrier transport in direct bandgap semiconductors, J Appl Phys 54: 1390-1398.

Würfel P (1995), Is an illuminated semiconductor far from thermodynamic equilibrium, Solar Energy Materials and Solar Cells 38: 23-28.

Würfel P, Finbeiner S and Daub B (1995), Generalized Planck's radiation law for luminescence via indirect transitions, Appl Phys A 60: 67-70.

Yablonovitch E (1982), Statistical ray optics, J Opt Soc Am 72: 399-907.

Yablonovitch E and Cody GD (1982), Intensity enhancement in textured optical sheets for solar cells, IEEE Trans on Electron Devices ED-29: 300-305.

5 Tandem Cells

5.1 Spectrum Splitting and Stacking

One of the major energy losses in single junction cells arises from the rapid thermalisation of photoexcited carriers particularly those with energy well in excess of the bandgap (process 1 of Fig. 4.1). This loss is associated with high entropy generation rate on light absorption (Sect. 4.10).

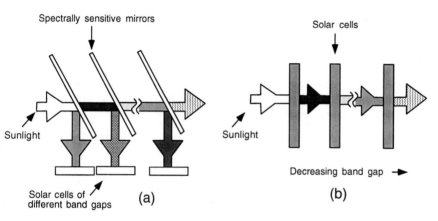

Fig. 5.1: Tandem cell concepts: (a) spectrum splitting; (b) cell stacking.

One way of reducing this loss is to subdivide the broad solar emission spectrum into different energy ranges and to convert each range with a cell of a well matched bandgap (Fig. 5.1). This idea seems to have been first suggested in 1955 (Jackson 1955), although not attracting much interest at the time (if multiple cells were to be used, it was thought the optimal use would be to have them working side by side, all producing as much power output as possible!). However, the first experimental demonstration of the concept in 1978 (Moon et al. 1978), and the high efficiencies obtained, re-awakened interest in the concept. The initial focus was upon concentrating systems where the improved efficiency would be an advantage in offsetting the cost of the associated lenses and sun tracking mechanisms.

It was realised right from the start that spectral filtering could be achieved automatically, merely by stacking cells on top of one another, with the largest bandgap cell uppermost (Jackson 1955). Early experimental work (Moon et al. 1978), however, was based on the spectrum-splitting approach of Fig. 5.1(a). An important development was the realisation that this approach may not only be applicable to expensive concentrator cells, but also had benefits at the other extreme of potentially low cost thin-film amorphous silicon cells (Hamakawa et al. 1981). In this case, stacking cells allowed the use of thinner cells, improving carrier collection prospects even in very low quality material.

Three-cell tandems became commercially available in the latter half of the 1990s in the form of high performance cells for spacecraft (Karam et al. 1998) and also, at the other extreme, of potentially low cost terrestrial thin-film cells based on amorphous Si:Ge:H alloys (Yang et al. 1998). Evolution to 4-cell tandems for space applications seems likely, with experimental efficiency above 40% also likely. There is less incentive for increasing the number of stacked cells for terrestrial use due to the increasing sensitivity to the spectral content of light.

5.2 Split-Spectrum Cells

In the spectrum splitting approach of Fig. 5.1(a), each cell operates in isolation and can be analysed by a simple extension of the approach applied to single junction devices. Assuming the mirrors are ideal in that they direct light onto each cell ranging from that cell's bandgap, E_{Gn}, to that of the next highest bandgap cell, $E_{G(n+1)}$, the ideal current-voltage relationship for each cell is given by:

$$I_n / (qA) = f_s \dot{N}(E_{Gn}, E_{G(n+1)}, 0, T_s) + (f_c - f_s)\dot{N}(E_{Gn}, E_{G(n+1)}, 0, T_c)$$
$$- f_c \dot{N}(E_{Gn}, E_{G(n+1)}, qV_n, T_c) \tag{5.1}$$

If all cells operate independently, the operating voltage V_n for each cell is optimised for maximum output (independent operation is cumbersome in practice due to the large number of individual connections required with other possibilities discussed in Sect. 5.4).

One feature that differs from the single junction case concerns the light emission from the cell at high energies. For the former case, with a 6000 K solar spectrum, photons are available at high energy so the optimum design is to have a very large energy upper limit E_M on light absorption and hence emission from the cell. However, in the present case, there is no need to absorb these photons because the next cell will, and also no need to emit them, so the optimum value of $E_M = E_{G(n+1)}$. This spectral sensitivity can be achieved, in principle, by placing a low energy pass filter in front of each cell that reflects all the photons of energy higher than those assigned to it. (For the configuration of Fig. 5.1(a), a horizontal mirror could be placed above each of the spectrally sensitive reflectors shown, such as shown in Fig. 3.4. This would divert this high energy light back to the cell emitting it).

The limiting efficiencies that result from this formulation for various numbers of cells are shown as the "diffuse, unconstrained" and "direct, unconstrained" entries of Table 5.1.

Table 5.1: Limiting efficiencies and optimal bandgaps for a range of tandem cell designs (Marti and Araujo 1996; Brown 2002).

No. of Cells	Description	Optimal Bandgaps (eV)						Effic %
		E1	E2	E3	E4	E5	E6	
1	Diffuse	1.31						31.0
	Direct	1.11						40.8
2	Diffuse, series connected	0.97	1.70					42.5
	Diffuse, unconstrained	0.98	1.87					42.9
	Direct, series connected	0.77	1.55					55.5
	Direct, unconstrained	0.77	1.70					55.9
3	Diffuse, series connected	0.82	1.30	1.95				48.6
	Diffuse, unconstrained	0.82	1.44	2.26				49.3
	Direct, series connected	0.61	1.15	1.82				63.2
	Direct, unconstrained	0.62	1.26	2.10				63.8
4	Diffuse, series connected	0.72	1.10	1.53	2.14			52.5
	Diffuse, unconstrained	0.72	1.21	1.77	2.55			53.3
	Direct, series connected	0.51	0.94	1.39	2.02			67.9
	Direct, unconstrained	0.52	1.03	1.61	2.41			68.8
5	Diffuse, series connected	0.66	0.97	1.30	1.70	2.29		55.1
	Diffuse, unconstrained	0.66	1.07	1.50	2.03	2.79		56.0
	Direct, series connected	0.44	0.81	1.16	1.58	2.18		71.1
	Direct, unconstrained	0.45	0.88	1.34	1.88	2.66		72.0
6	Diffuse, series connected	0.61	0.89	1.16	1.46	1.84	2.41	57.0
	Diffuse, unconstrained	0.61	0.96	1.33	1.74	2.26	3.00	58.0
	Direct, series connected	0.38	0.71	1.01	1.33	1.72	2.31	73.4
	Direct, unconstrained	0.40	0.78	1.17	1.60	2.12	2.87	74.4
∞	Diffuse (unconstrained, series connected, 2-terminal)							68.2
	Direct (unconstrained, series connected, 2-terminal)							86.8

5.3 Stacked Cells

In the case of stacked cells, there are more opportunities for interaction between the cells. Fig. 5.2 shows the simplest case. A cell in the middle of the stack receives sunlight that has had high-energy photons removed by preceding cells. It also emits light from both its surfaces and receives light emitted by cells on either side. There is clearly a certain economy in such recycling of photons.

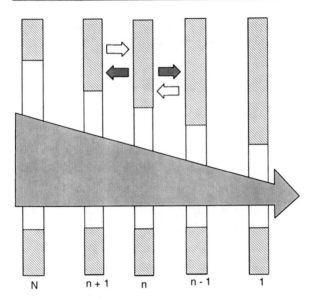

Fig. 5.2: Stacked tandem solar cells. Radiation emitted by cell *n* from its rear is converted by cell *(n – 1)* while some of that emitted from its front is converted by cell *(n + 1)*. After (Marti and Araujo 1996).

Recycling introduces additional terms into the equation governing the performance of each cell, coupling its performance to those of the cells on either side (Marti and Araujo 1996). However, it has been shown that the more isolated mode of operation, as discussed in Sect. 5.2, also leads to marginally higher performance. This mode of operation can be implemented, in principle, in stacked cells by inserting wavelength dependent filters between cells as indicated in Fig. 5.3. In practice, the performance gain is so marginal (only a few percentage points at most) that the extra complexity would not be warranted. However, the structure is consistent with finding the highest possible performance from the approach. It also has the advantage of being analytically simpler.

As the number of cells becomes large, the small difference between the two configurations becomes even smaller, so both approach the same limiting efficiency value as the number of cells approaches infinity (Marti and Araujo 1996).

A third configuration is also possible where adjacent cells are joined by semiconductor material of high refractive index *n*. This is the case of most practical interest as discussed in the following section. Although it is possible to contemplate building low energy pass filters into such semiconductor regions, this is not often attempted. Coupling between the cells is increased by the factor n^2, reducing achievable efficiency (Tobias and Luque 2002) for finite, but not infinite, cell numbers. An advantage of this case is a higher tolerance to cell mismatch. Current output in cells deep in the stack can be boosted by radiative transmission from overlying cells (Brown and Green, 2002a).

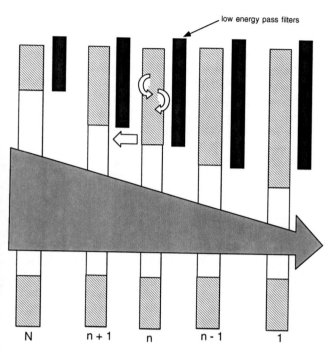

Fig. 5.3: Wavelength dependent filters placed between stacked cells provide, in principle, a marginal performance increase. These filters prevent emission of light to lower bandgap cells, which more than compensates for the consequent disadvantage of not being able to emit high energy photons to higher bandgap cells (Marti and Araujo 1996).

5.4 Two Terminal Operation

The benefits of increased power output would be offset by increased system complexity if separate connections to individual cells had to be made. Tandem cells are generally operated as two-terminal devices although other options are discussed elsewhere (Frass 1995). For crystalline III-V compound cells, "monolithic" two terminal technology is generally used whereby cells are grown epitaxially on top of the same substrate, with interconnection provided by a tunnelling junction between each pair of cells.

For such series-connected cells, matching of the current outputs of the cells is essential, as for standard solar modules (Wenham et al. 1994). This matching is achieved by controlling the bandgap and thickness of each cell in the stack. As shown in Table 5.1, the constraint imposed by such series connections slightly reduces the obtainable efficiency.

A more severe consequence for terrestrial cells is the increased sensitivity to the spectral content of light. Once the output of a cell in a series connection drops more than about 5% below that of the next best cell, the low output cell can severely disrupt the power generating capabilities of the connection. As a number of cells increases, each with an increasingly narrow section of the solar spectrum to convert, larger fluctuations in the current output of constituent cells will become likely with changes of the solar spectrum. This would be expected to provide a practical limit on the maximum number of cells in the stack.

As noted in the previous section, monolithic structures increase the coupling between the cells and can, in some situations, reduce the sensitivity to spectral mismatch (Brown and Green 2002a). As the number of cells in the stack approaches infinity (following section), it can be proved that the efficiency of the two terminal configuration approaches that of the separately contacted cells, regardless of the possible optical couplings between the cells (Brown and Green 2002b; Tobias and Luque 2002).

5.5 Infinite Number of Cells

The case of an infinite number of cells is interesting since it allows the upper limit on the efficiency of the tandem approach to be determined. More generally, it also allows insight into the fundamental constraints on photovoltaic conversion.

For an infinite number of cells, each cell converts photons over an infinitesimal energy range, dE, with a current-voltage relationship given by:

$$I = \frac{2\pi q A}{h^3 c^2}\left[\frac{f_s E^2}{exp(E/kT_s)-1} + \frac{(f_c - f_s)E^2}{exp(E/kT_c)-1} - \frac{f_c E^2}{exp[(E-qV)/kT_c]-1}\right]dE \quad (5.2)$$

The optimum operating voltage can be found by multiplying by V and differentiating to find the equation for the voltage, V_m, that maximises this output. This gives:

$$\frac{f_s}{exp(E/kT_s)-1} + \frac{(f_c - f_s)}{exp(E/kT_c)-1} - \frac{f_c\{(1+qV_m/kT_c)exp[(E-qV_m)/kT_c]-1\}}{\{exp[(E-qV_m)/kT_c]-1\}^2} = \quad (5.3)$$

The optimum efficiency of the stack is then given by:

$$\eta = \frac{15q/f_s}{\pi^4 k^4 T_s^4}\int_0^\infty\left[\frac{f_s V_m E^2}{exp(E/kT_s)-1} + \frac{(f_c - f_s)V_m E^2}{exp(E/kT_c)-1} - \frac{f_c V_m E^2}{exp[(E-qV_m)/kT_c]-1}\right]dE \quad (5.4)$$

Solving Eq. (5.3) for each cell in the stack (each value of $E = hf$) will allow the optimum operating voltage of each cell and hence the optimum efficiency of the stack to be found. For the case of $f_s = f_c$, Eq. (5.3) simplifies to:

$$V_m = E(1-T_c/T_s\,)/q-(kT/q\,)ln\left\{\frac{1+(qV_m/kT\,)-exp[(qV_m-E\,)/kT\,]}{1+[(qV_m/kT\,)-1]\,exp[(qV_m-E\,)/kT\,]}\right\}$$

(5.5)

The first term on the right represents the open-circuit voltage, given by a Carnot relationship in this case, and the second term represents a small voltage offset that guarantees the overall conversion efficiency is less than the Carnot efficiency.

5.6 Approximate Solution

The optimum performance can be estimated using intuition about solar cell operation. For example, one strategy that has been used for tracking the maximum power point voltage of solar cells is to operate at a voltage that is a fixed function of the open-circuit voltage (Stephens 1990). This compensates well for changes in cell temperature and light intensity. Assume this strategy also works well for changes in the bandgap. For a direct conversion cell ($f_c = f_s$), the open-circuit voltage of each cell is a fixed fraction $(1 - T_c/T_s)$ of the value hf/q. Hence, this strategy would give an operating voltage for each cell equal to Fhf/q, where F is a constant.

The Bose Einstein function determining the light emission from each cell would then become:

$$f_{BE} = \frac{1}{e^{hf(1-F)/kT_c}-1} = \frac{1}{e^{hf/kT_e}-1}$$

(5.6)

where T_e is an effective temperature equal to $T_c/(1 - F)$. Substituting back into Eq. (5.2) for cell current and noting that the cell voltage equals $Fhf/q = (1 - T_c/T_e)hf/q$, the cell efficiency becomes:

$$\eta = (1-T_e^4/T_s^4\,)(1-T_c/T_e\,)$$

(5.7)

This is an equation that has surfaced before in Chap. 3 as Eq. (3.10). Its solution is $T_e = 2544K$ and $\eta = 85.4\%$. This corresponds to operating each cell at a voltage equal to 92.9% of its V_{oc}. The calculated efficiency is close to the value of 86.8% obtained with the full calculation, confirming the strength of this voltage tracking strategy.

For a cell operating under diffuse sunlight, this tracking strategy is non-optimal. The problem is that the strategy forces the low bandgap cells to emit more light than they absorb from the sun, with a lot of the sun's energy being converted to re-radiated infrared light.

A better strategy in this case would be just to immobilise these cells by short-circuiting or open-circuiting them or, better still, removing them entirely from the stack.

5.7 Thermodynamics of the Infinite Stack

As previously noted, for the direct light case $f_s = f_c$, on open circuit, each cell reaches an open-circuit voltage of $(1 - T_c/T_s)\, hf/q$. At open circuit, the amount of light emitted by each cell exactly balances that absorbed from the sun. This is the condition for zero entropy production. Infinitesimally close to open-circuit, heat supplied to fuel the solar emission would be converted to electricity by the cells at the Carnot efficiency (95.0%). This suggests the infinite tandem cell case is close to a thermodynamic ideal.

In practice, there is no credit for the light emitted back to the sun. The conversion efficiency of available sunlight to electricity approaches zero near open-circuit. The cell has to be biased to lower voltages to obtain useful power output as per Eq. (5.5). This causes an imbalance between the effective temperature of absorption and emission, resulting in entropy production.

Since, for each energy, the voltage of the infinite tandem can be adjusted for the best possible compromise, it turns out that the infinite tandem is an example of a photovoltaic scheme displaying the limiting photovoltaic performance. This makes the corresponding calculated efficiency limits of 86.8% and 68.2% for direct and diffuse cells the upper limit of attainable photovoltaic performance for such a "reciprocal" converter regardless of conversion approach.

Exercises

5.1 Write down the equation corresponding to Eq. (5.1) for the case of stacked tandem cells (without filters) as in Fig. 5.2.

5.2 Find the optimum operating voltage of the cells of the following bandgaps in an infinite stack of tandem cells capable of converting diffuse radiation, assuming $T_s = 6000$ K and $T_c = T_A = 300$ K: (a) 0.25 eV; (b) 0.5 eV; (c) 2.0 eV.

References

Brown AS and Green MA (2002a), Radiative coupling as a means to reduce spectral mismatch in monolithic tandem solar cell stacks – theoretical considerations, paper presented at 29th IEEE Photovoltaic Specialists Conference, May.

Brown AS and Green MA (2002b), Limiting efficiency for current-constrained two-terminal tandem cell stacks, Progress in Photovoltaics 10: 299-307.

Fraas LM (1995), Concentrator modules using multijunction cells in L. Partain (Ed.), "Solar Cells and Their Applications", Wiley, New York.

Hamakawa Y, Okamoto H, Nitta Y and Adachi T, Photovoltaic cell array having multiple vertical PIN junctions, US Patent 4,271,328, June 2, 1981 (Priority Date: 20 March, 1979).

Jackson ED (1955), Trans. Conf. on Use of Solar Energy, Tuscon, Arizona, Oct. 31 – Nov. 1, 122 (also US Patent 2,949,498, Aug. 16, 1960).

Karam NH, Ermer JH, King RR, Hadda M, Cai L, Joslinn DE, Krut DD, Takahashi M, Eldredge JW, Nishikawa JW, Cavicchi BT and Lillington DR (1998), High efficiency GaInP$_2$/GaAs/Ge dual and triple junction solar cells for space applications, Proceedings, 2nd World Conference on Photovoltaic Solar Energy Conversion, Vienna, 6-10 July, 3534-3539.

Marti A and Araujo GL (1996), Limiting efficiencies for photovoltaic energy conversion in multigap systems, Solar Energy Materials and Solar Cells 43: 203-222.

Moon RL, James LW, Vander Plas HA, Yep TO, Antypas GA and Chai Y (1978), Multigap solar cell requirements and the performance of AlGaAs and Si cells in concentrated sunlight, Conf Record, 13th IEEE Photovoltaic Specialists Conference, Washington, DC, 859-867.

Stephens A (1990), Temperature dependence of silicon solar cell performance, BE Thesis, University of New South Wales.

Tobias I and Luque A (2002), Ideal efficiency of monolithic, series-connected multijunction solar cells, Progress in Photovoltaics 10: 323-329.

Wenham SR, Green MA and Watt M (1994), Applied Photovoltaics, (Bridge Printery, Sydney) (available from author).

Yang J, Banerjee A, Lord K and Guha S (1998), Correlation of component cells with high efficiency amorphous silicon alloy triple-junction solar cells and modules, Proceedings, 2nd World Conference on Photovoltaic Solar Energy Conversion, Vienna, 6-10 July, 387-390.

6 Hot Carrier Cells

6.1 Introduction

One of the major loss processes for standard cells (Fig. 4.1) arises from the rapid thermalisation of the photoexcited carriers with the lattice. If this thermalisation can be avoided, much higher efficiency would be possible in principle. In this case, carriers could still thermalise by collisions with one another to stabilise in a distribution described by a temperature much higher than the lattice temperature (Fig. 6.1). This chapter explores the design features required for hot carrier operation and the efficiencies that can be obtained in limiting cases.

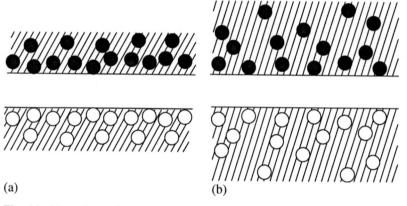

(a)   (b)

Fig. 6.1: Normally carriers thermalise with the lattice as in (a). In hot carrier cells, excess energy is stored in a hot carrier distribution, as shown in (b).

6.2 Relevant Time Constants

An insight into the different time constants involved in photoexcitation and subsequent relaxation processes can be gained by considering the case of a semiconductor illuminated by a short, high intensity pulse of light, such as from a laser.

Figure 6.2 shows the evolution with time of electron and hole distributions in a semiconductor subject to such pulsed excitation.

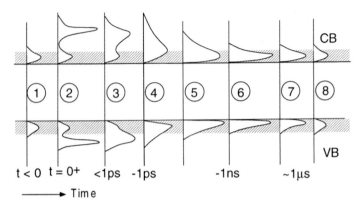

Fig. 6.2: Time evolution of electron and hole distributions in a semiconductor subject to a short, high intensity, monochromatic pulse of light from a laser: (1) Thermal equilibrium before pulse; (2) "coherent" stage straight after pulse; (3) carrier scattering; (4) thermalisation of "hot carriers"; (5) carrier cooling; (6) lattice thermalised carriers; (7) recombination of carriers; (8) return to thermal equilibrium.

Before the laser pulse, the semiconductor is in thermal equilibrium. Electron and hole concentrations in the conduction and valence band, respectively, are described by a Boltzmann distribution, decaying exponentially with increasing energy from the band edge, but modulated by the increasing density of states available to them as this energy increases. This gives the characteristically peaked functions as this energy increases as shown (the peak ideally occurs $0.5 \, kT$ in energy beyond the band edge under non-degenerate conditions).

Immediately after laser pulsing, the additional photogenerated carrier population is superimposed upon this equilibrium concentration. For monochromatic laser light, these carriers are excited from and to specific regions of *energy–momentum* space, giving sharply peaked "coherent" distributions as shown. The peak in the valence band will generally lie closer to the band edge than that in the conduction band, due to the generally higher valence band effective mass. This means that more of the surplus photon energy is absorbed in the conduction band than in the valence band.

The laser excitation now terminates and the carriers begin their long process back towards equilibrium. Initially, carrier-carrier collisions come into effect and act to more uniformly distribute the excess energy amongst the carriers. This is the type of process analysed by Ludwig Boltzmann in the 1860s. The carriers evolve towards a Boltzmann distribution in energy, as in thermal equilibrium, but one characterised by a much higher temperature than that at thermal equilibrium.

During this carrier collision stage, there is no energy lost; energy is just more democratically shared between the photoexcited and original carriers. The effective temperature of the carriers will depend upon the total energy shared amongst them. Electrons and holes could initially reach distributions characterised by different temperatures, depending upon the amount of electron-hole scattering taking place (Othonos 1998).

In the next phase, the hot carriers start to lose energy to the lattice by phonon emission. Initially this loss will be via the emission of high energy optical phonons but eventually it will be dominated by lower energy acoustic phonons. During this phase, the total number of electrons and holes remains virtually unchanged. However, their average energy and "temperature" decreases as they progressively lose energy.

Eventually, the carriers cool down sufficiently so that their temperature matches that of the lattice. The next dominant phase in their relaxation involves the recombination of excess electron-hole pairs. This progressively reduces the number of excess carriers until equilibrium is re-established.

A conventional solar cell operates with carrier collection times of the same order as the latter recombination process. A well designed cell will collect most of the carriers after they thermalise with the lattice, but before they have a chance to recombine. A hot carrier cell has to work on much shorter time scales. It has to collect the carriers before they start to cool from their high temperature self-thermalised state.

Since carriers have a finite velocity, this means there will be a limit on the distance they are able to travel to be collected prior to cooling. Hot carrier thermalisation distances are typically of the order of 10nm or so, giving an indication of possible collection distances.

Another important consideration is the way the device is contacted. Since carriers within the contact are thermalised at the lattice temperature, they must be prevented from vigorous interaction with the hot carriers in the cell. If such interaction occurs, the contacts will cool the hot carriers. This interaction can be minimised by withdrawing carriers across only a vanishingly small range of energies, rather than across the entire conduction or valence band as in a conventional cell.

One suggestion of how this might be done (Würfel 1997) is by use of a wide bandgap semiconductor with narrow conduction and valence bands (Fig. 6.3).

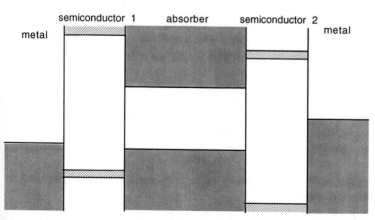

Fig. 6.3: Selective energy contacts to hot carrier cell based on wide bandgap semiconductors with narrow valence (left) and conduction (right) bands.

Another possibility is to use resonant tunnelling devices which ideally allow the passage of carriers with longitudinal energy matching that of the energy level in the intermediate well. Figure 6.4 shows a possible resonant tunnelling scheme.

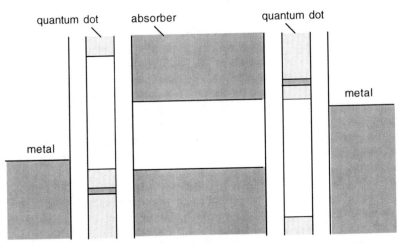

quantum dot   absorber   quantum dot

metal

metal

Fig. 6.4: Possible energy selective contacting scheme based on resonant tunnelling. Ideally, carriers tunnel between the absorber and contacts at an energy corresponding to that of the bound state in the semiconductor quantum well.

6.3 Ross And Nozik's Analysis

The first systematic analysis of hot carrier cells (Ross and Nozik 1982) was stimulated by a suggestion that hot carrier cells might be expected to display similar efficiencies to an infinite tandem cell (Archer 1981). Ross and Nozik assumed that carriers in the conduction and valence band came to the same equilibrium temperature, T_H, and that the respective distributions could be described by separate quasi-Fermi energies μ_C and μ_V with their difference given by:

$$\Delta\mu_H = \mu_C - \mu_V \tag{6.1}$$

To minimise interaction with the external environment, they considered that carriers were removed from one specific energy in the conduction band and returned at one specific energy in the valence band (Fig. 6.4), separated by an energy ΔE_{use}.

The same particle balance argument as in previous chapters can be used to deduce the extracted current:

$$I_{use}/(qA) = f_s\dot{N}(E_G,\infty,0,T_s) + (f_c - f_s)\dot{N}(E_G,\infty,0,T_A) - f_c\dot{N}(E_G,\infty,\Delta\mu_H,T_H) \tag{6.2}$$

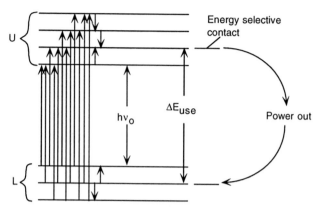

Fig. 6.5: Schematic of hot carrier cell showing carrier extraction and return (Ross and Nozik 1982).

Since, unlike the standard cell, there is no energy loss in a hot carrier device in the ideal limit, an energy balance can also be applied:

$$E_{use}I_{use} / A = f_s \dot{E}(E_G,\infty,0,T_s)+(f_c - f_s)\dot{E}(E_G,\infty,0,T_A)- f_c \dot{E}(E_G,\infty,\Delta\mu_H,T_H)$$
(6.3)

The power output from the cell equals the extracted current times the potential difference between the contacts. This does not equal $\Delta\mu_H/q$ since the contacts are at ambient temperature and $\Delta\mu_H$ is a parameter describing the carrier distributions at a higher temperature. The corresponding electrochemical potential difference in an ambient temperature distribution needs to be found (Fig. 6.6).

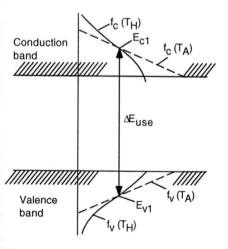

Fig. 6.6: Energy distribution functions for carriers within a hot carrier cell and corresponding ambient temperature distributions that match at the extraction and return energies, E_{C1} and E_{V1}, respectively.

Equating the distributions at E_{Cl} and E_{Vl} gives:

$$(E_{C1} - \mu_{CH})/T_H = (E_{C1} - \mu_{CA})/T_A \qquad (6.4)$$

$$(E_{V1} - \mu_{VH})/T_H = (E_{V1} - \mu_{VA})/T_A \qquad (6.5)$$

Subtracting gives:

$$\Delta E_{use} / T_H - \Delta \mu_H / T_H = \Delta E_{use} / T_A - \Delta \mu_A / T_A \qquad (6.6)$$

or

$$\Delta \mu_A = \Delta \mu_H T_A / T_H + \Delta E_{use}(1 - T_A / T_H) \qquad (6.7)$$

The power output then becomes:

$$P_{use} = I_{use} \Delta \mu_A / q$$
$$= (I_{use} / q) \Delta \mu_H T_A / T_H + \Delta E_{use}(I_{use} / q)(1 - T_A / T_H) \qquad (6.8)$$

The first term can be expressed in terms of particle flows from Eq. (6.2) and the second term in terms of energy flows from Eq. (6.3). From Eq. (6.8), P_{use} can be calculated as a function of E_G, T_H and $\Delta \mu_H$ (Ross and Nozik 1982). Interestingly, good performance can be obtained even if E_G approaches zero. One way of thinking of this is that finite E_G provides a "safety-net" in a normal cell that temporarily delays the energy relaxation of the photoexcited carriers. In a hot carrier cell, such a safety net is not required since carriers are collected before they have a chance to lose energy.

Figure 6.7 (curves labelled $\Delta_{\mu H} < 0$)) shows the calculated limiting efficiency as a function of bandgap for the conversion of both direct and diffuse light. Efficiencies quite close to the tandem limit are possible. The values of T_H required to reach this limiting performance are quite high, about 4000 K for the conversion of direct sunlight and 3900 K for the conversion of diffuse sunlight, although performance close to optimal is obtained over a wide range of carrier temperatures (anything above 1500 K gives reasonably good results). Note that quite different values of $\Delta \mu_H$ are required for optimal performance when converting direct and diffuse sunlight. For direct sunlight, $\Delta \mu_H \approx -0.06\ T_H/T_A$ (eV), while for diffuse sunlight $\Delta \mu_H \approx -0.3\ T_H/T_A$ (eV). The negative value of $\Delta \mu_H$ in both cases helps to improve efficiency by suppressing light emission.

Equation (6.8) contains a number of simpler situations as limiting cases. For example, with $T_H = T_A$, the second term becomes zero and the analysis reverts to that of the standard single junction cell of Chap. 4. With $\Delta \mu_H = 0$, the first term drops out. The second terms assumes a simple form for $E_g = 0$ corresponding to an efficiency equal to (neglecting sky radiation):

$$\eta = [1 - T_H^4 / (f_s T_s^4 / f_c)](1 - T_A / T_H) \qquad (6.9)$$

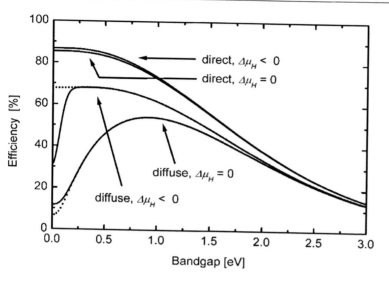

Fig. 6.7: Limiting efficiency of a hot carrier cell for direct and diffuse sunlight. The curves labelled $\Delta\mu_H < 0$ show the unconstrained case while the curves labelled $\Delta\mu_H = 0$ shows the case where there are high levels of interaction between hot electrons and holes.

This form also occurred in Chaps. 3 and 5 in different contexts. For $f_s = f_c$, this expression gives an efficiency of 85.4% for $T_H = 2544K$. As apparent in Fig. 6.7, the condition $\Delta\mu_H = 0$ gives quite close to the optimum for the conversion of direct light, although far removed from the optimum for the conversion of diffuse light.

The reason why the efficiency of the hot carrier conversion process is slightly lower than the infinite tandem case relates to the fact that there are insufficient variables available to ensure that the conversion at each photon energy is occurring at the optimal chemical potential. Generalising the arguments leading to Eq. (6.7), it is seen that the chemical potential referenced to ambient temperature for each photon energy is given by:

$$\Delta\mu = \Delta\mu_H T_A / T_H + hf(1 - T_A / T_H)$$ (6.10)

Essentially, there are two independent parameters T_H and $\Delta\mu_H$ that can be adjusted to give a good, but not perfect fit to the optimum values of $\Delta\mu$ as determined by solving Eq. (5.3) As seen in Fig. 6.8(a), quite a close fit to the optimum can be found. The situation is much improved over the case of a single junction where there is little flexibility in fitting to this optimum. With $\Delta\mu_H = 0$, quite a good fit to the optimum can be maintained for the direct light case, as seen in Fig. 6.8(b), although the fit is very poor for the case of diffuse light.

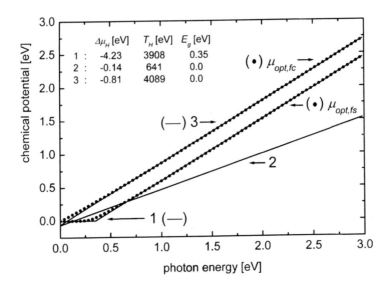

(a)

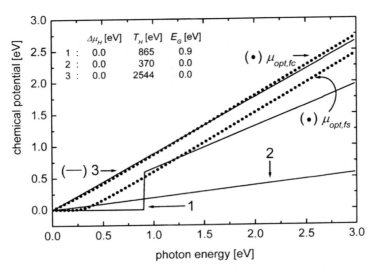

(b)

Fig. 6.8: Optimal chemical potential versus photon energy obtained by solving Eq. (5.3) shown as dotted lines for the case where $f_S = f_C$ (uppermost dotted curve on both graphs) and $f_C = 1$ (lowermost dotted curve on both). The other curves correspond to the linear approximations possible in the two hot carrier approaches for the conditions shown in the inset: (a) Ross and Nozik's analysis ($\Delta\mu_H < 0$ allowed) and (b) Würfel's analysis ($\Delta\mu_H = 0$).

6.4 Simplification for $E_g = 0$

Equation (6.8) can be simplified for the case where $E_G = 0$ and where $\Delta\mu_H/kT_H$ equals a reasonably large negative quantity. This is an important case since, from Fig. 6.7, it corresponds to the region of maximum energy conversion efficiency. Such negative values of $\Delta\mu_H/kT$ allow the following approximation to be used:

$$\int_0^\infty \frac{E^j dE}{exp[(E-\Delta\mu_H)/kT_H]-1} \approx exp(\Delta\mu_H/kT_H)\int_0^\infty \frac{E^j dE}{exp(E/kT_H)} \quad (6.11)$$

Inserting in to Eq. (6.8) gives:

$$\eta = \frac{\Delta\mu_H}{2.701kT_H}\left(\frac{T_c}{T_s}\right)\left[1+\left(\frac{f_c-f_s}{f_s}\right)\left(\frac{T_c}{T_s}\right)^3-\frac{(f_c/f_s)}{1.202}\left(\frac{T_H}{T_s}\right)^3 exp(\Delta\mu_H/kT_H)\right]$$

$$+\left(1-\frac{T_c}{T_H}\right)\left[1+\frac{(f_c-f_s)}{f_s}\left(\frac{T_c}{T_s}\right)^4-\frac{(f_c/f_s)}{1.082}\left(\frac{T_H}{T_s}\right)^4 exp(\Delta\mu_H/kT_H)\right]$$

$$(6.12)$$

where the numbers appearing equal $[\Gamma(4)\xi(4)/(\Gamma(3)\xi(3))]$, $\xi(3)$ and $\xi(4)$, respectively.

This equation is readily solved for each T_H to find the optimum $\Delta\mu_H$, and hence the maximum possible efficiency from the approach.

6.5 Würfel's Analysis

Recently, a more specific situation has been analysed (Würfel 1997) that gives additional insight into hot carrier operation as well as providing a link to the analysis of the cells that generate multiple electron-hole pairs per photon, as described in the next chapter.

Würfel analyses the case where again there is no loss of energy by the carriers by phonon emission, but there is strong interaction between electrons and holes not only by scattering but also by impact ionisation and by the reverse process of Auger recombination.

Würfel imagined a specific case whereby energy is conserved in a reaction where $dN_{eh,1}$ electron-hole pairs of energy $\varepsilon_{eh,1}$ are transformed to $dN_{eh,2}$ electron-hole pairs of energy $\varepsilon_{eh,2}$ such that:

$$dN_{eh,1}\varepsilon_{eh,1} + dN_{eh,2}\varepsilon_{eh,2} = 0 \quad (6.13)$$

This models impact ionisation and Auger recombination processes. In equilibrium, the free energy associated with such interactions is at a minimum giving additionally:

$$dN_{eh,1}\mu_{eh,1} + dN_{eh,2}\mu_{eh,2} = 0 \tag{6.14}$$

where μ_{eh} is the chemical potential of the pairs. Combining Eqs. (5.13) and (6.14) gives:

$$\mu_{eh,2} = \frac{\varepsilon_{eh,2}}{\varepsilon_{eh,1}} \cdot \mu_{eh,1} \tag{6.15}$$

or, more generally, the chemical potential of electron-hole pairs is proportional to their energy:

$$\mu_{eh} = \beta\varepsilon_{eh} = \beta(\varepsilon_e + \varepsilon_h) \tag{6.16}$$

Scattering of electrons and holes also puts constraints on equilibrium distributions. If two electrons in states 1 and 2 are scattered to give two electrons in states 3 and 4, minimisation of the free energy imposes the constraint:

$$dN_{e,1}\mu_{e,1} + dN_{e,2}\mu_{e,2} + dN_{e,3}\mu_{e,3} + dN_{e,4}\mu_{e,4} = 0 \tag{6.17}$$

Noting that $dN_{e,1}\mu_{e,1} = dN_{e,2}\mu_{e,2} = -dN_{e,3}\mu_{e,3} = -dN_{e,4}\mu_{e,4}$, gives:

$$\mu_{e,1} + \mu_{e,2} = \mu_{e,3} + \mu_{e,4} \tag{6.18}$$

This is compatible with all electrons having the same chemical potential. However, it is also compatible with energy dependent Fermi-levels of the type:

$$\mu_{e,i} - \mu_{e,0} = \beta\varepsilon_{e,i} \qquad \mu_{h,i} - \mu_{h,0} = \beta\varepsilon_{h,i} \tag{6.19}$$

Where $\mu_{e,0}$ and $\mu_{h,0}$ are reference levels. Equation (6.19) agrees with Eq. (6.16) if $\mu_{e,0} + \mu_{h,0} = 0$.

Note that, for example, the electron energy distribution would be described by:

$$dN_e(\varepsilon_e) = \frac{D(\varepsilon_e)d\varepsilon_e}{exp[(\varepsilon_e - \mu_e(\varepsilon_e))/kT] + 1} \tag{6.20}$$

Where $D(\varepsilon_e)$ is the density of states for electrons. The term in square brackets can be simplified by invoking Eq. (6.19) giving:

$$dN_e(\varepsilon_e) = \frac{D(\varepsilon_e)d\varepsilon_e}{exp[(\varepsilon_e - \mu_{e,0})(1-\beta)/kT] + 1} \tag{6.21}$$

This shows that the energy distribution of electrons in the conduction band can be described either by energy-dependent electrochemical potentials at temperature, T, or by a single electrochemical potential valid for all energies and an effective temperature $T_{eh} = T/(1 - \beta)$.

Having established that a hot-carrier distribution with $\Delta\mu_H = 0$ is characteristic of strongly interactive electron-hole pairs in the presence of impact ionization and Auger recombination (but in the absence of phonon scattering), Würfel's analysis then follows the earlier analysis of Ross and Nozik. Würfel particularly stresses the need for isoentropic cooling of the carriers at the contact, suggesting the contacting scheme of Fig. 6.3, previously described.

The efficiencies calculated using Würfel's analysis are shown as the curves labelled $\Delta\mu_H = 0$ in Fig. 6.7. For $E_g = 0$, they can be described by Eq. (6.9). The advantages of negative values of $\Delta\mu_H$ are modest for the conversion of direct sunlight but quite marked for diffuse sunlight.

6.6 Possible Low Dimensional Implementation

It was seen earlier that resonant tunnelling diodes may provide the required energy selectivity in making contact to hot carrier devices (Fig. 6.4). Also, the dimensions of the absorber have to be small to allow carriers to be collected prior to thermalisation with the lattice. This also suggests low dimensional structures might be especially well suited for use as hot carrier devices.

An additional effect that has been observed is reduced carrier cooling rates in multiple quantum well and superlattice devices when excited to high carrier injection levels (Arent et al. 1993). This is attributed to "hot" phonon populations under such conditions that reduce the effectiveness of phonons in the cooling process (the hot phonons share their energy with the carriers). Unsuccessful experimental attempts to use this effect to produce hot carrier cells have been reported (Hanna et al. 1997), but without the use of the energy selective contacts that are likely to be necessary for hot carrier operation.

A very active area of optoelectronics concerns "photonic" engineering, where 3-D modulation of optical properties of semiconductor devices is being used to control permissible optical propagation energies. Low dimensional structures allow the modulation of mechanical properties that may allow "phononic" engineering to control phonon emission and carrier cooling effects.

Very thin absorber regions are likely to be required to accommodate the short collection distances. Hot carrier equivalents of multilayer cell designs might be required, where layer thickness is small compared to the thermalisation distance. Alternatively, technology similar to that used in nanocrystalline dye cells could be used, where a very convoluted surface structure allows strong absorption even when the absorber layer is very thin (Ernst el al. 2000). To approach the limiting efficiencies described, collection times must not only be very sort compared to energy relaxation times, but radiative recombination times must also be faster, implying very strongly absorbing structures.

Exercise

6.1   Derive Eq. (6.12) from Eq. (6.8). For the conversion of direct sunlight, find the optimum $\Delta\mu_H$ and conversion efficiency for $T_H = 4000\ K$. Repeat this calculation for the conversion of diffuse sunlight for $T_H = 3600$ K (assume $T_C = 6000$ K and $T_s = T_A = 300$ K).

References

Archer MD (1981) in Photochemical Conversion and Storage of Solar Energy, Academic Press, New York, J.S. Connolly (Ed.), 333.

Arent DJ, Szmyd D, Hanna MC, Jones KM and Nozik AJ (1993), Hot electron cooling in parabolic and modulation doped quantum wells and doped superlattices, Superlattices and Microstructures 13: 459-468.

Ernst K, Lux-Steiner MC and Konenkamp R (2000), All-solid state and inorganic solar cell with extremely thin absorber based on CdTe, paper presented at 16th European Photovoltaic Solar Energy Conference, Glasgow, 1-5 May, 63-66.

Hanna MC Lu Z and Nozik AJ (1997), Hot carrier solar cells, Proceedings, 1st NREL Conference on Future Photovoltaic Generation Technologies.

Othonos A (1998), Probing ultrafast carrier and phonon dynamics in semi-conductors, J. Appl Phys 83: 1789-1830.

Ross RT and Nozik AJ (1982), Efficiency of hot-carrier solar energy converters, J Appl Phys 53: 3813-3818.

Würfel P (1997), Solar energy conversion with hot electrons from impact ionisation, Solar Energy Materials and Solar Cells 46: 43-52.

7 Multiple Electron-Hole Pairs per Photon

7.1 Introduction

The energy of high-energy photons can be used to improve solar cell energy conversion efficiency by creating multiple electron-hole pairs per incident photon. This is quite an old idea (Deb and Saha 1972) that has recently resurfaced (Kolodinski et al. 1993, 1994). Performance limits have been analysed recently by an approach consistent with that of earlier chapters (Werner et al. 1995).

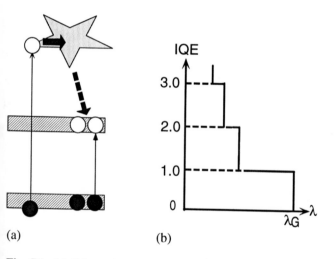

(a)   (b)

Fig. 7.1: (a) Schematic of the impact ionisation process whereby one energetic photon creates multiple electron-hole pairs; (b) Energetically feasible internal quantum efficiency, where λ_G is the wavelength corresponding to the threshold energy for electron-hole pair creation.

The basic concept is shown in Fig. 7.1(a). If a high energy photon creates an electron-hole pair with most of its excess energy concentrated in one band, it is possible for this hot carrier to give up its excess energy by a collision with a lattice atom, creating a second electron-hole pair in the process. Such impact ionisation becomes energetically possible once the photon energy is greater than twice the band gap energy.

However, there is no reason to stop here! Once the photon energy becomes greater than three times the bandgap, three electron-hole pairs are energetically feasible. In principle, the internal quantum efficiency of the cell could approach that of Fig. 7.1(b). Clearly, greatly enhanced efficiency is a possibility in principle. In practice to date, gains have been extremely modest with the highest observed internal quantum efficiency for the best silicon cells being less than 1.05 at a wavelength of 350nm, compared to the limiting value of 3.0 (Wilkinson et al. 1983), increasing to 1.5 at 250 nm, where a value of 4.0 is energetically feasible (Wolf et al. 1996). The case for other semiconductors appears no more promising (Alig et al. 1980; Landsberg et al. 1993).

In this chapter, the limiting efficiency for this process is determined, involving some unexpected subtleties. Also, other processes apart from impact ionisation are suggested as offering more potential for obtaining the predicted performance advantages.

7.2 Multiple-Carrier Photon Emission

Initial thinking was that the bonus of the improved spectral sensitivity would be obtained without any resulting sacrifice. The corresponding formulation would proceed as for a standard single junction device but with the solar generated current suitably augmented. However, calculations based on this approach showed unphysical features under extreme conditions. For example, efficiencies higher than the Carnot efficiency were predicted under some conditions (Werner et al. 1994).

This indicated there was something wrong with this apparently reasonable formulation. What was missing was incorporation of the physical fact that, if a photon can be absorbed creating multiple electron-hole pairs, then multiple electron-hole pairs could recombine to create a single photon. Such a photon would be associated with a different chemical potential from that associated with photons generated by the recombination of single electron-hole pairs.

For example, if m electrons recombine with m holes to produce a single photon, the equilibrium free energy argument shows that:

$$m\mu_e dN_e + m\mu_h dN_\lambda + \mu_{ph} dN_{ph} = 0 \qquad (7.1)$$

Hence, $\mu_{ph} = m\mu_{eh}$. If the generation of multiple electron hole pairs per incident photon is permitted, the emitted photon population is augmented by photons associated with different multiples of the electron-hole chemical potential. A thermodynamic analysis of this aspect is presented elsewhere (Luque and Marti 1997).

A more physical description of this aspect is that single electron-hole processes are proportional to the number of these pairs and hence to $exp\ (qV/kT)$ in the Shockley-Queisser analyses of Sect. 4.2. Processes involving two of these pairs will increase as the square of their concentration and hence as $exp\ (2qV/kT)$. Processes involving m pairs will increase as $exp\ (mqV/kT)$.

7.3 Limiting Performance

With this variable chemical potential included into the formulation, the output current of the cell is given by, for the ideal case of Fig. 7.1(b):

$$I = \frac{2\pi q A}{h^3 c^2} \sum_{m=1}^{\infty} \int_{mEG}^{(m+1)EG} [\frac{mf_s}{exp(\,E\,/\,kT_s\,)-1} + \frac{m(\,f_c - f_s\,)}{exp(\,E\,/\,kT_A\,)-1}$$
$$- \frac{mf_c}{exp[(\,E - mqV\,)/\,kT_c\,]-1}]E^2 dE$$

(7.2)

where the factor, m, occurs in the first two terms due to multiple electron-hole pair generation and in the third term since the recombination of multiple electron-hole pairs is involved in the emission of a single photon. Multiplying I by V gives the power output. Identifying the staircase function mqV with the chemical potential $\mu(E)$, the power output becomes:

$$P = \frac{2\pi A}{h^3 c^2} \int_{EG}^{\infty} (\frac{f_s \mu(\,E\,)}{exp(\,E/kTs\,)-1} + \frac{(\,f_c - f_s\,)\mu(\,E\,)}{exp(\,E/kT_A\,)-1}$$
$$- \frac{f_c \mu(\,E\,)}{exp[(\,E - \mu(\,E\,))/kT_c\,]-1})E^2 dE$$

(7.3)

This is now in the familiar form also applicable to other conversion options discussed in previous chapters. For maximum conversion efficiency, $\mu(E)$ must approximate the optimal value given by Eq. (5.3) as closely as possible. Figure 7.2 is a schematic showing the trends in calculated efficiency limits with this parameter optimised (solid lines). Also shown is a schematic of the results from the analysis of a similar situation by Würfel, described in the previous chapter (Würfel 1997) .

A simplified case where analytical solution is possible is where E_g approaches zero and the staircase function $\mu(E)$ approaches the value aE/E_g. In this case, the integrals can be evaluated analytically leading to the now familiar result:

$$\eta = \{1 - (\,T_c\,/\,T_s\,)^4 + (\,f_c\,/\,f_s\,)[(\,T_c\,/\,T_s\,)^4 - (\,T_I\,/\,T_s\,)^4\,]\}(1 - T_c\,/\,T_I\,) \quad (7.4)$$

This gives a peak of 85.4% efficiency for $f_s = f_c$. A detailed study near $E_g = 0$ shows that this does not represent the optimal efficiency (de Vos and Besoete 1998). The latter occurs for a small but finite bandgap of 0.048 eV, giving a peak efficiency of 85.9%. As indicated in Fig. 7.3, this gives a slightly better match to the optimal chemical potential variation with energy, by providing a slight offset near $hf = 0$. For diffuse light conversion, optimal conversion occurs for a bandgap of 0.76 eV, giving a peak efficiency of 43.6%. This is somewhat lower than for the ideal diffuse conversion efficiency limit for some of the other high efficiency options investigated, due to the relatively coarse match to the optimal chemical potential variation with photon energy, as indicated by Fig. 7.3.

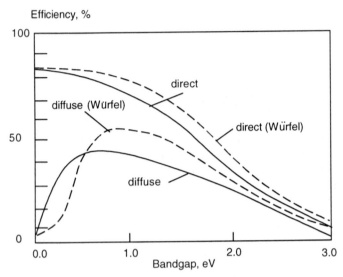

Fig. 7.2: Schematic (only!) showing the approximate value of the limiting efficiency for solar cells where multiple electron-hole pairs are generated per incident photon, up to the number that is energetically feasible (solid line). Also represented are the calculations by Würfel (dashed lines).

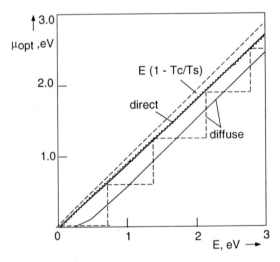

Fig. 7.3: Optimal chemical potential as a function of photon energy for direct and diffuse light (continuous lines) and the best matches possible with a staircase function (schematic only).

7.4 Comparison with Würfel's Analysis

Both Würfel's analysis (Sect. 6.5) and the previous analysis seek to describe the same physical situation but give the same results only at $E_G = 0$ (Fig. 7.2). Interestingly, Würfel's analysis gives better results for some values of E_G. This is despite the fact that it sets out to be more conservative, including presumably detrimental Auger recombination, the inverse of impact ionisation, into the analysis (Würfel 1997).

The difference relates to the treatment of phonon interaction. Würfel found that, to get a performance boost in the case he analyses where there is very strong interaction between electrons and holes, he needed to assume negligible interaction with phonons. This results essentially in "hot carrier" operation of the cell. However, the analysis of Sect. 7.3 neglects Auger recombination as a competing inverse process, but assumes the carriers are not hot, but thermalise with the lattice.

Würfel's analysis assumes each photon is created as a result of a recombination event involving a single electron and a hole. The photon emission spectrum therefore reflects the hot carrier distribution. The analysis of Sect. 7.3 predicts an unusual photon emission spectrum that approaches that from a hot carrier distribution as E_G approaches zero. This type of emission (typical of a hot carrier distribution) occurs despite the fact that carriers have a standard "cool" lattice-thermalised distribution. The unusual photon distribution arises from the fact that multiple electron-hole pairs are assumed to recombine to give a single photon.

Würfel argues that low levels of interaction with phonons are essential if impact ionisation is to provide any performance gain. Such a conservative conclusion would seem to be at least partly supported by the experimental facts. For example, for the case of silicon, it has been noted in Sect. 7.1 that impact ionisation provides negligible performance boost. However, Auger recombination prevents silicon from reaching its radiative performance limit even for diffuse light. It is even more restrictive for direct light (Green 1984). Increasing impact ionisation in silicon cells might be expected to also increase the inverse Auger recombination rates, further reducing performance below the radiative ideal. Even GaAs cells can be prevented from reaching their radiative limit for direct light conversion by Auger recombination [6.14], indicating the generality of the problem identified.

However, it is possible to conceive of approaches that would improve impact ionisation effectiveness without increasing Auger recombination rates. For example, if competing energy loss processes for excited electrons such as by phonon emission were reduced in magnitude, multi-pair production rates would improve without any effect upon Auger recombination rates.

Additionally, as discussed in Sect. 7.6, other processes apart from impact ionisation can result in multiple electron-hole pair creation per incident photon, removing Auger recombination as a fundamental issue.

7.5 Possible Implementation

As previously mentioned, the experimentally measured boost from impact ionisation is too modest to be of any practical significance. Competing processes for the relaxation of energetic carriers are too efficient. Figure 7.4 shows the type of excitation possible that conserves both energy and crystal momentum when the primary excitation is at the zone centre. Since silicon conduction and valence bands are parallel over large volumes of energy momentum space, there are many other options for such processes once photon energy exceeds the direct bandgap value of 3.4 eV. Manipulation of the bandgap, such as by alloying in surface regions with Ge, could lower this threshold (Kolodinski et al. 1995). Band folding effects in Si/Ge or Si/SiO$_2$ superlattices could do the same. Since the high energy photons involved in multiple electron-hole pair generation are likely to be strongly absorbed, such modifications may be required only in the cell surface region.

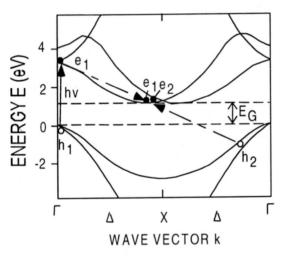

Fig. 7.4: Energy band diagram of silicon showing ain impact ionisation event conserving both energy and crystal momentum (Kolodinski et al. 1994).

7.6 Generalised Analysis

A more general, less idealised analysis of multiple electron-hole pair creation is possible by defining absorption coefficients, α_m, for processes involving the generation of m electron-hole pairs. For $m = 0$, this would be the free carrier absorption coefficient. For $m = 1$, this corresponds to the usual band-to-band absorption processes, with α_1 having reasonably large values for energies larger than the semiconductor bandgap but finite values even below the bandgap due to phonon assisted processes (Green 1995). The case of $m = 2$ corresponds to absorption creating two electron pairs and would not be expected to be significant except for energies larger than twice the semiconductor bandgap. α_m corresponds to processes creating m pairs and is only likely to be important for energies greater than m times the bandgap.

There will be a limit on m determined by issues such as photo-ionisation to an upper limit k. The normally measured absorption coefficient at any energy will just be the sum of these values:

$$\alpha_{tot} = \sum_{m=0}^{k} \alpha_m \tag{7.5}$$

Equation (4.35) can be generalised to:

$$u. \nabla f_{pt} = \sum_{m=0}^{k} \alpha_m [f_{BE}(m\mu_{CV}) - f_{pt}] \tag{7.6}$$

For a thick device, this will result in an equilibrium photon state occupancy given by:

$$f_{pt}^{eq} = \sum_{m=0}^{k} \alpha_m f_{BE}(m\mu_{CV}) / \sum_{m=0}^{k} \alpha_m = \sum_{m=0}^{\infty} \alpha_m f_{BE}(m\mu_{CV}) / \alpha_{tot} \tag{7.7}$$

Following the analysis of Sect. 4.6, the diode current can be expressed as:

$$I = \frac{2\pi q A}{h^3 c^2} \left\{ f_s \int_0^\infty \frac{\left(\sum_{m=0}^{k} m\alpha_m / \alpha_{tot}\right) E^2 dE}{e^{E/kT_s} - 1} + (f_C - f_s) \int_0^\infty \frac{\left(\sum_{m=0}^{k} m\alpha_m / \alpha_{tot}\right) E^2 dE}{e^{E/kT_C} - 1} \right.$$

$$- f_C \int_0^\infty [\sum_{m=0}^{k} m\alpha_m / \alpha_{tot}] f_{pt}^{eq} E^2 dE$$

$$\left. - \int^{Vol.} \int_0^\infty (4n^2 / \alpha_{tot}) \sum_{m=1}^{k} \sum_{j=0}^{m-1} \alpha_m \alpha_j (m-j)[f_{BE}(mqV) - f_{BE}(jqV)] E^2 dEdVol / A \right\}$$

$$\tag{7.8}$$

The first two terms represent photon absorption from the source taking into account that the process represented by absorption coefficient α_m generates m electron-hole pairs. The third term represents a combination of the current flow needed to support photon emission, taking into account the increased current when multiple electron-hole pairs recombine to form a single photon, plus a correction from the fourth and final term. The fourth term represents the additional current needed to support the reabsorption of photons involving one or more electron-hole pairs by processes involving fewer electron-hole pairs, uncorrected for the fact that some of the internally generated photons are emitted. A description of these terms in a simpler case is given in Sect. 4.6.

How important are these multiple pair generation effects in actual solar cells? For silicon, it is of interest to note that the 2 pair process shown in Fig. 7.4 is direct, not requiring phonons. The first transition at the Γ point could also be a virtual transition, allowing such phononless transitions once the energy equalled twice the bandgap. One might think a process involving a photon and two valence band electrons might not be much weaker than one involving a photon, a valence band electron, plus a phonon. However, at twice the bandgap (550 nm wavelength) and even higher energies (wavelengths to 400 nm), there is no evidence for multiple electron-hole pair creation within the normal accuracy of internal quantum efficiency measurements (about 1%). This suggests that, for silicon, $\alpha_2 < 0.01\ \alpha_1$ over this wavelength range. At 350 nm, the internal quantum efficiency is less than 1.05 suggesting that, at this wavelength:

$$2\ \alpha_2 + 3\ \alpha_3 < 0.05\ \alpha_1$$

At this wavelength and shorter, there would be many points in the Brillioun zone where a direct photon excitation could occur, giving good prospects for the creating of 2 and 3 electron hole-pairs, still without involving phonons. Clear evidence for the predicted photon emission hump at twice the bandgap energy with twice the chemical potential has been observed in germanium at room temperature (Condradt and Waidelich 1968).

If it were possible to improve the effectiveness of impact ionisation in creating multiple electron-hole pairs without dramatically changing other material properties, such as by clinically removing only the strongest of the competing processes, we might not expect to change the measured silicon absorption coefficient at these energies significantly. What would be expected to change is the partitioning between α_1 and α_2, with α_1 reducing to compensate the increased value of α_2.

The generalised Eq. (7.5) suggests that α_{m+1}/α_m has to be large for best results from this approach. Since, for standard semiconductors, α_m is not likely to decrease with increasing energy, this means that the ideal material would have a step increase in its absorption coefficient at each multiple of its bandgap. Both silicon and germanium have such step changes as their absorption changes from indirect to direct, but this step does not occur at twice the bandgap energy in either case. Alloying these two would allow this condition to be met (Kolodinski et al. 1995).

7.7 Raman Luminescence

Other processes apart from impact ionisation offer additional prospects for implementing multiple pair generation. One such process is Raman luminescence as shown in Fig. 7.5(a). Here an incoming energetic photon creates an electron-

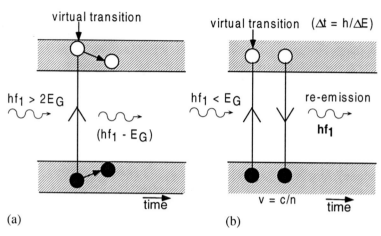

Fig. 7.5: (a) Raman luminescence; (b) related virtual absorption process for sub-bandgap photons.

hole pair, with any additional energy emitted as a second photon. (This process is similar to the Raman scattering processes used in material characterisation, where the second photon differs in energy from the first by the energy of a phonon involved in the scattering processes).

The likely strength of such processes can be investigated by comparing with the virtual excitation processes that determine the high refractive index of semiconductors at sub-bandgap wavelengths (Fig. 7.5(b)). Although photons of such wavelengths do not have sufficient energy to create an electron-hole pair, their passage through the semiconductor can be described in terms of a series of virtual transitions. The low energy photon is involved in the successive virtual excitation of an electron-hole pair, essentially borrowing energy via the uncertainty principle. The virtual pair can remain viable for a time determined by this principle $(\Delta t \approx h/\Delta E)$. If nothing happens to resolve this transition over this period, such as the absorption of another photon or of multiple phonons to provide an energy balance, the virtual pair will disappear, emitting a photon of identical energy to that involved in the original transition but somewhat delayed. This delay slows the passage of light through the material to the value c/n where n is here the material refractive index.

Interestingly, for silicon, the virtual transitions accounting for its high refractive index are direct band-to-band transitions, generally requiring photon energies in excess of 3.4eV if realised as actual transitions. The contributions to the refractive index by these virtual transitions can be determined by second order perturbation theory as (Klingshirm 1995):

$$[n(hf)^2 - 1] \propto \left| \sum_z \frac{<i|H_{eR}|z><z|H_{eR}|i>}{(E_z - E_i - hf)} \right|^2 \tag{7.9}$$

where the sum is over all possible intermediate sates, z, and H_{eR} represents the strength of the interaction between electrons and radiation. The normal Raman scattering, involving phonons, is also formally described in terms of such virtual transitions. The probability of such scattering is given by the sum of six terms of the following form (Yu and Cardona 1996):

$$P = \frac{(2\pi)^2}{h} \left| \sum_{z,z'} \frac{<i|H_{eR}(hf')|z'><z'|H_{el}|z><z|H_{eR}(hf)|i>}{(E_{z'} - E_i - hf')(E_z - E_i - hf)} \right|^2 \qquad (7.10)$$

Raman luminescence bears a formal similarity to 2-photon absorption, since the same cast of particles is involved. The strength of the Raman luminescence is given by:

$$P = \frac{(2\pi)^2}{h} \left| \sum_z \frac{<f|H_{eR}(hf')|z><z|H_{eR}(hf)|i>}{(E_z - E_i - hf)} \right|^2 \qquad (7.11)$$

This is quite similar in form to Eq. (7.9). It would appear to be a more probable process than that described by Eq. (7.10), since fewer terms are involved. For a strong presence, the transition matrix for excitations from the ground to the intermediate state should be large, as also should be that between the intermediate and final state. To ensure large values for these elements, these transitions should be between states of opposite parity (e.g., between states of an s-like character and those of a p-type character). From this, it follows that the initial and final states should have the same parity.

How else can Raman luminescence be enhanced compared to competing processes? In the process shown in Fig. 7.5(a), but with wide bands, the luminescence process would have to compete with direct absorption followed by relaxation of the carriers by phonon emission or even impact ionisation. The direct absorption process could be discouraged by choosing material where both conduction and valence bands were of finite width. Once the photon energy exceeded the bandgap plus the sum of these bandwidths, only virtual transitions would be possible for at least a limited range of energies, clearly enhancing their relative strength!

With such constrained bandwidths, Raman luminescence could be enhanced by virtual transitions to higher bands (Fig. 7.6(a)), particularly as resonance (energy conservation) was approached during the virtual transition.

As previously noted, the relative strength of the Raman luminescence would be enhanced if the virtual transitions were "allowed" transitions, while the transitions to the lowest conduction band were "forbidden". Many reasonably common semiconductors have a direct bandgap but the transitions between the uppermost valence and conduction band is "dipole" forbidden since both bands have the same parity (when calculating the transition probabilities between these bands, the similar symmetry of these bands results in zero net probability near the zone centre). Examples are Cu_2O, SnO_2, TiO_2 and GeO_2 (Klingshirm 1995).

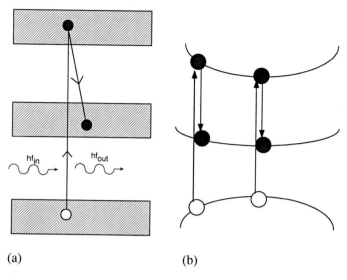

(a)   (b)

Fig. 7.6(a) and (b): Use of near-resonant Raman scattering to enhance scattering rates. Real as well as virtual transitions between bands can also contribute to the conversion process.

Transitions between the bands are still possible but their strength is very much reduced by the symmetries involved. If transitions to the next highest conduction band were "allowed", this would act to enhance the relative strength of Raman luminescence (Raman luminescence would involve two "allowed" transitions, competing with direct transitions, involving one "forbidden").

In crystalline material, wave vector must be conserved during both "virtual" and "real" transitions (Fig. 7.6(b)). This constraint may help increase design opportunities. "Flat" bands (high effective mass and low carrier mobilities) seem most conducive to enhancing Raman luminescence.

Transitions to higher lying bands could be "real" with subsequent radiative relaxation of carriers back to lower lying bands. This case would then resemble a special operating mode of the multiband cells described in Chap. 8.

Low dimensional structures also provide other possibilities for enhanced Raman luminescence. For example, Fig. 7.7 shows a superlattice structure with corresponding minibands in the conduction and valence bands. In the longitudinal direction, this structure displays energy-momentum relationships similar to those of Fig. 7.6(b). In the transverse direction, however, these bands can overlap, undesirably enhancing transitions between them by non-radiative processes. A 3-D superlattice based on quantum dots rather than quantum wells would remove this deficiency, but poses additional experimental challenges.

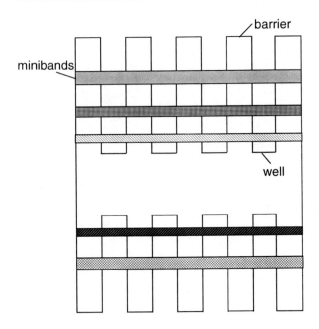

Fig. 7.7: Superlattice bands may enhance near-resonance Raman luminescence for transitions either between valence and conduction band minibands or between minibands associated with either the conduction or valence band.

Exercises

7.1 Derive Eq. (7.5) for a solar cell in the case of multiple electron-hole pair generation per incident photon up to the limit that is energetically feasible, as the semiconductor bandgap approaches zero.

7.2 From the information in the text, estimate the relative magnitudes of the emitted photon flux spectral values at the bandgap and twice the bandgap for silicon at 0.7 volts bias, taking into account emission involving two electron-hole pairs.

References

Alig RC, Bloom S and Struck CW (1980), Scattering by ionization and phonon emission in semiconductors, Phys Rev B 22: 5565-5582.

Araujo GL and Marti A (1990), Limiting efficiencies GaAs solar cells, IEEE Transactions on Electron Devices 37: 1402-1405.

Conradt R and Waidelich W (1968), Indirect band-to-band Auger recombination in Ge, Physical Review Letters 20: 8-9.

de Vos A and Besoete B (1998), On the ideal performance of solar cells with larger-than-unity quantum efficiency, Solar Energy Materials and Solar Cells 51: 413-424.

Deb S and Saha H (1972), Secondary ionisation and its possible bearing on the performance of a solar cell, Solid State Electronics 15: 89-1391.

Green MA (1984), Limits on the open circuit voltage and efficiency of silicon solar cells imposed by intrinsic Auger processes, IEEE Trans Electron Devices ED-31: 671-678.

Klingshirm CF (1995), Semiconductor Optics, Springer-Verlag, Berlin, 192.

Kolodinski S, Werner JH, Wittchen T and Queisser HJ (1993), Quantum efficiencies exceeding unity due to impact ionization in silicon solar cells, Applied Physics Letters 63: 2405-2407.

Kolodinski S, Werner JH, Wittchen JH and Queisser HJ (1994), Quantum efficiencies exceeding unity in silicon leading to novel selection principles for solar cell materials, Solar Energy Materials and Solar Cells 33: 275-286.

Kolodinski S, Werner JH and Queisser HJ (1995), Potential of $Si_{1-x}Ge_x$ alloys for Auger generation in highly efficient solar cells, Applied Physics A 61: 535-539.

Landsberg PT, Nussbaumer H and Willeke G (1993), Band-band impact ionization and solar cell efficiency, J Appl Phys 74: 1451-1452.

Luque A and Marti A (1997), Entropy production in photovoltaic conversion, Physical Review B 55: 6994-6999.

Werner JH, Brendel R and Queisser HJ (1994), New upper efficiency limits for semiconductor solar cells, Conference Record, IEEE First World Conference on Photovoltaic Energy Conversion, Hawaii, 1742-1745.

Werner JH, Brendel R and Queisser HJ (1995), Radiative efficiency limit of terrestrial solar cells with internal carrier multiplication, Applied Physics Letters 67: 1028-1030.

Wilkinson FJ, Farmer AJD and Geist J (1983), The near ultraviolet yield of silicon, Journal of Applied Physics 54: 1172-1174.

Wolf M, Brendel R and Werner JH (1996), Quantum efficiency of silicon solar cells at low temperatures, Tech Digest of the International PVSEC-9, Miyazaki, Japan, 519-520.

Würfel P (1997), Solar energy conversion with hot electrons from impact ionisation, Solar Energy Materials and Solar Cells 46: 43-52.

Yu PY and Cardona M (1996), Fundamentals of Semiconductors, Springer-Verlag, Berlin, 379-385.

8 Impurity Photovoltaic and Multiband Cells

8.1 Introduction

One idea for efficiency improvement suggested some time ago was to extend cell response by excitations through energy states due to defects, lying at energies within the normally forbidden gap of the host semiconductor (Wolf 1960). This allows the cell to respond to photons of energy below the bandgap (Fig. 8.1). An undesirable consequence is that the presence of the defect levels also enhances recombination in the cell, at one time thought to negate any benefits (Guttler and Queisser 1970). Subsequent analyses of both specific (Keevers and Green 1994) and generic defects (Würfel 1993) concluded that only minor performance improvements might be expected from this approach. More recent work using the ideas outlined in previous chapters has shown a very clear advantage for devices if able to operate at close to radiative limits (Brown and Green 2002a).

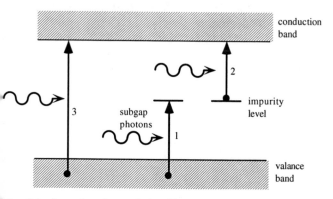

Fig. 8.1: Impurity photovoltaic effect where electron-hole pairs are generated by sub-bandgap photons (Keevers and Green 1994).

A more modern version of this idea is the use of multiple quantum well solar cells (Barnham and Duggan 1990). In this case, excitations in the lower bandgap well regions (Fig. 8.2) provides a different energy threshold from excitations across the wider bandgap barrier regions. There has been considerable discussion as to whether or not this approach offers the potential for any performance improvement (Araujo and Marti 1995). It now seems there is the potential for performance improvement if the "thermal escape" mechanisms of Fig. 8.2 are

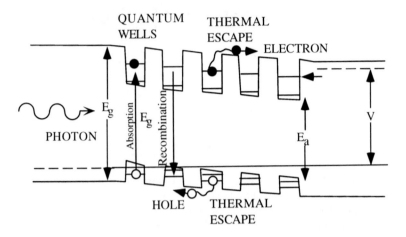

Fig. 8.2: Multiple quantum well solar cell (Barnham and Duggan 1990).

replaced by processes mediated by particles introducing additional chemical potential, e.g., solar photons or "hot" phonons (Luque et al. 2001).

Considerable insight into the potential of such approaches has been gained by a recent analysis of the limiting performance of a multiband solar cell (Luque and Marti 1997). In the case analysed (Fig. 8.3), sufficient impurities were considered to have been added to the host semiconductor for their normally isolated energy levels associated with them to interact sufficiently strongly to form a conductive impurity band (Milne 1973; Shklovskii and Efros 1994). This assumption allows the Shockley-Queisser detailed balance approach to be extended to this case,

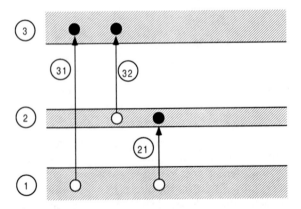

Fig. 8.3: 3-band solar cell. The lower- and upper-most bands are valence and conduction bands, while the intermediate band is considered to be an impurity band.

allowing a general assessment of its efficiency potential as well as suggesting options for its extension

8.2 3-Band Cell

A central assumption of the analysis is that high energy photons do not waste their energy on excitations possible with lower photon energy (e.g., in Fig. 8.3, photons with sufficient energy for process 31 are not involved in lower energy excitations such as process 21). Approaches to achieving this are discussed in Sect. 8.3. If such photon absorption selectivity can be ensured, it is possible to write down the by now familiar particle balance for each transition:

$$I_{hl} = qA[f_s \dot{N}(E_l, E_h, 0, T_s) + (f_c - f_s) \dot{N}(E_l, E_h, 0, T_A) - f_c \dot{N}(E_l, E_h, \mu_{hl}, T_c)]$$

(8.1)

where I_{hl} is the current flowing between the two bands being considered, E_l and E_h are the lower and upper photon energies involved in the corresponding transition, and μ_{hl} is the chemical potential or quasi-Fermi energy difference between these bands. This equation applies in turn for each of the transitions shown in Fig. 8.3. Since the middle band is not contacted, the current flowing in and of this band must match in the steady-state ($I_{12} = I_{23}$). Also, since carriers in each band are assumed to have thermalised with one another, but not with carriers in other bands, each band is associated with a single chemical potential. It follows that:

$$\mu_{31} = \mu_3 - \mu_1 = \mu_3 - \mu_2 + \mu_2 - \mu_1 = \mu_{32} + \mu_{21}$$

(8.2)

These constraints on current and chemical potential can be best represented in terms of the equivalent circuit of Fig. 8.4(a). Solving the relevant equations subject to these constraints for the case where $f_s = f_c$ gives an optimum efficiency of 63.2% for the case where the lowest threshold energy is 0.7 eV, the next lowest is 1.2 eV and the highest is 1.9 eV (Luque and Marti 1997). The above efficiency is very close to the value of 63.7% which is the corresponding limit on an unconstrained 3-cell tandem (Table 5.1). It is almost identical to the limit on a series-connected 3-cell tandem, whose equivalent circuit is shown in Fig. 8.4(b), also calculated as 63.2% (Brown and Green 2002). Figure 8.5 shows the calculated limiting performance as a function of the threshold energy for the two lower energy processes for the case of small intermediate band width.

There are some advantages compared to a 3-cell series-connected tandem, however. One is that the same semiconductor material is used throughout and interconnection between the sub-units occurs automatically in the 3-band case. Another is reduced spectral sensitivity. The three band cell is over twice as robust to changes in spectral content as a 3-cell series connected tandem device.

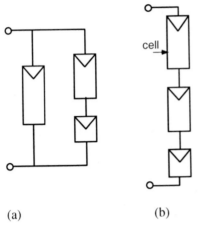

(a)   (b)

Fig. 8.4: (a) Equivalent circuit of a 3-band cell; (b) Equivalent circuit of a monolithic 3-cell tandem (the length of the cell representation indicates its threshold energy).

8.3 Photon Absorption Selectivity

To reach the limiting performance of Fig. 8.5, high energy photons cannot be wasted on low energy processes. Any such misappropriation or leakage represents an energy loss that obviously will detract from overall efficiency. Approaches for avoiding this loss are outlined in the following sections.

8.3.1 Finite Bandwidths

One way of ensuring the required photon selectivity is if the width of all the bands involved in the cell are finite and reasonably small. This means there would automatically be an upper and lower limit on the energy of the photons eligible for each of the possible excitation processes (Green 1999, 2000).

For example, for the 3-band cell, the optimal design (Fig. 8.6) is to have the middle band quite narrow, as would be the case if it were an impurity band. It would ideally be separated from one of the bands, say the valence band, by 0.7 eV and from the other band by 1.2 eV, giving the optimal separation between the lowest and highest lying bands of 1.9 eV. The valence band would ideally be 0.5 eV wide (minus the width of the impurity band) so that photons of 0.7 eV to 1.2 eV energy could excite electrons from the valence to the impurity band. Similarly, the conduction band would be 0.7 eV wide so that photons of 1.2 eV to 1.9 eV energy could excite them from the impurity to the conduction band. Photons of 1.9 eV to 3.1 eV energy would then be able to excite electrons from the valence to the conduction band.

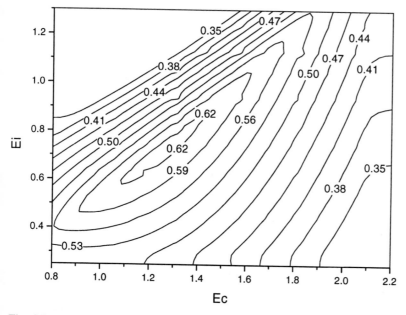

Fig. 8.5: Limiting efficiency of a 3-band cell as a function of the two lower threshold energies (Corkish 1999). $(T_s = 6000 \text{ K}, T_c = 300 \text{ K})$.

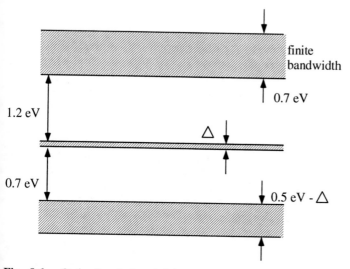

Fig. 8.6: Optimally designed 3-band cell with photon selectivity ensured by finite bandwidth for each of the 3 bands involved.

Note that a disadvantage of this approach to photon selectivity is that there is a cap on the maximum energy of a photon able to participate (in this case, 3.1 eV).

This reduces attainable efficiency from 63.2% to 57.1%, for the optimal design (Corkish 1999). In practice, this would not be as severe a limitation as this result implies since high energy photons are generally filtered out of sunlight before it strikes the cell, either by ozone in the earth's atmosphere or by ultraviolet absorbers deliberately added to encapsulants such as glass and potants, to improve the ultraviolet stability of the latter. Additional modelling of the absorption properties of the design of Fig. 8.6 has been reported elsewhere (Cuadra et al. 2000).

8.3.2 Graded Absorption Coefficients

A second approach to ensuring photon selectivity is to have a gradation in absorption properties so that the highest energy processes are the most strongly absorbing and the lowest energy, the most weakly absorbing. This will ensure that high energy photons are preferentially absorbed in those processes able to utilise them (Green 1999).

A recent analysis of this strategy (Cuadra et al. 2000) shows that a grading ratio of absorption coefficients above 2 will produce better limiting cell performance than for a single junction cell (40.8%), while values in the 10-100 range produce efficiencies quite close to the limiting efficiency (63.2%).

8.3.3 Spatial Absorption Partitioning

A third approach to photon selectivity is based on spatial partitioning of the absorption processes (Green 1999). By designing the cell so that the incoming light is exposed first to a region only able to absorb high energy photons, these high energy photons can be filtered from the incoming light. The light then passes to a region only able to absorb high and intermediate energy photons, and finally to a region able to absorb these and low energy photons. Three examples are shown in Fig. 8.7. This spatial approach ensures optimal use is made of incoming light but does not necessarily guarantee this for recycled photons.

The scheme in Fig. 8.7(a) relies on an impurity band to form the intermediate band, although this band is not continuous through the whole device (Green 1999). High energy photons are absorbed near the cell surface in band-to-band processes, since the impurity band is absent. Doping is used to control the occupancy of this impurity band so that, at the point where light first encounters it, it is full of electrons and only excitations of these electrons to the conduction band are feasible. At the rear of the device, doping is again used to ensure the impurity band is largely depleted of electrons, enhancing transitions to it from the valence band. The effect of impurity band occupancy upon the relevant transition rates has recently been analysed (Cuadra et al. 2000), confirming this predicted behaviour.

Although impurity bands formed by high levels of dopants are well documented, the difficulty with this scheme may relate to problems with forming bands from deeper lying states, due to limited solubilities of the relevant impurities, combined with more tightly bound electron states. Optimal impurity band properties will be obtained if these impurities are regularly, rather than randomly, arranged within the host material.

The second approach in Fig. 8.7 (Green 1999) uses an interface between a wide bandgap and a narrow bandgap semiconductor, with carriers able to be optically excited over this interface. This latter light absorption process is likely to be weak compared to the other two. A high interfacial area such as demonstrated in nanocrystalline dye cells (Grätzel 2000) would help improve the strength of this process. However, another problem would be that not all carriers photoexcited near this interface would head over the barrier in the direction indicated. A better system may be the nanocrystalline dye system itself, where a multiband approach has recently been suggested (Grätzel 2000).

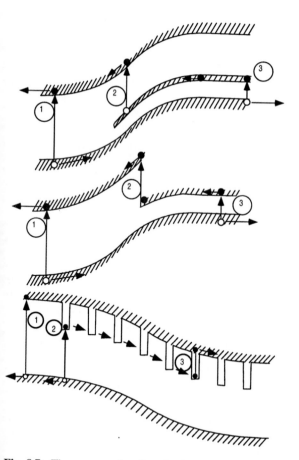

Fig. 8.7: Three approaches for selective photon absorption: (a) impurity band; (b) hetero-junction interface; (c) multiple quantum well.

Figure 8.7(c) shows a scheme based on multiple quantum wells (Green 1999). Absorption is again controlled spatially by the doping of the host semiconductor. To work in the 3-band mode, hopping of carriers between states in adjacent quantum wells has to be more efficient than escape from the well. Such quantum well hopping has been used to advantage in high gain photodetectors (Capasso et al. 1985). The ability to formulate the multiple quantum well approach in terms of 3-band theory clearly demonstrates that the quantum well approach offers a performance advantage over single junction devices in some specific cases.

Although it can be ensured that incoming light is absorbed by the appropriate process in the schemes of Fig. 8.7, this is not the case for recycled photons. Graded absorption coefficients can be used to improve the performance with regard to such recycled light. For the schemes of Fig. 8.7(a) and (c), high energy photons will be generated more or less uniformly throughout the cell volume by conduction to valence band recombination. Reabsorption in the inverse process will be favoured by the higher absorption strength of such a process. Some leakage to lower energy processes, as discussed in the next section will occur, however. Moderate energy photons will be generated in the central region of the device. These will suffer a fate similar to those emitted by central cells in an unfiltered stacked tandem cell (Fig. 5.2). Gains by conversion through high energy absorption process will be more than offset by losses from conversion in lower energy processes. Since there is no air gap between the different regions in this case, losses will be enhanced by a factor n^2, where n is the cell refractive index, as for monolithic tandem cells.

The case of Fig. 8.7(b) is a little different in that the generation regions for the different phonon energies are more compartmentalised. Performance similar to the lossy tandem case is expected for all energies.

8.4 Absorption Leakage Loss

The diversion of photons to non-optimal absorption processes results in a performance loss that has recently been analysed (Luque et al. 2000) and is related to the free-carrier absorption loss discussed in Sect. 4.6. Our starting point will be Eq. (4.36) where competition between two absorption processes has already been analysed. The obvious extension of this equation to the three level case becomes (Luque and Marti 1997; Luque et al. 2000):

$$\boldsymbol{u}.\nabla f_{pt} = \alpha_{31}[\, f_{BE}(\mu_{31}) - f_{pt}\,] + \alpha_{32}[\, f_{BE}(\mu_{32}) - f_{pt}\,] + \alpha_{21}[\, f_{BE}(\mu_{21}) - f_{pt}\,]$$

$$(8.3)$$

If the case is examined where we follow a light ray propagating in the direction $\boldsymbol{u}$, the left hand side reduces to df_{pt}/du and the solution for f_{pt} along the path of the ray becomes:

$$f_{pt} = f_{pt}(0)e^{-u\Sigma\alpha} + f_{pt}(\infty)(1 - e^{-u\Sigma\alpha})$$

$$(8.4)$$

where $f_{pt}(0)$ is the value at the surface, as discussed in Sect. 4.8, and $f_{pt}(\infty)$ corresponds to the spatially constant value, in this case given by:

$$f_{pt}(\infty) = [\alpha_{31} f_{BE}(\mu_{31}) + \alpha_{32} f_{BE}(\mu_{32}) + \alpha_{21} f_{BE}(\mu_{21})] / \Sigma\alpha \tag{8.5}$$

The consequences of this solution have recently been examined for the case of a planar cell, as in Fig. 4.6.

8.4.1 Current Equations

In the radiative limit, the total number of electrons flowing between any two bands is equal to the difference between the total absorption of photons exciting electrons between the bands throughout the device volume minus the total emission of photons by radiative relaxation of an electron back to the originating band. If only one absorption process is possible for any given photon energy, efficient photon recycling is possible. Emission and absorption are balanced through most of the cell volume, apart from the imbalance needed to absorb the solar photons and to provide for the enhanced thermal emission from the cell.

If competing absorption processes are present, there need no longer be this delicate balance between absorption and emission. Processes emitting large numbers of photons due to high chemical potential are no longer assured of recouping them. This creates an imbalance throughout the cell volume and hence a volume dependent contribution to cell currents (Luque et al. 2000).

By the above argument, Eq. (8.1) is modified to become:

$$
\begin{aligned}
I_{hl} = \frac{2\pi q A}{h^3 c^2} \{ f_s \int_{El}^{Eh} \frac{(\alpha_{hl} / \Sigma\alpha) E^2 dE}{e^{E/kT_s} - 1} + (f_c - f_s) \int_{El}^{Eh} \frac{(\alpha_{hl} / \Sigma\alpha) E^2 dE}{e^{E/kT_A} - 1} \\
- f_c \int_{El}^{Eh} (\alpha_{hl} / \Sigma\alpha) f_{pt}(\infty) E^2 dE - \int^{Vol} \int_{El}^{Eh} 4n^2 \alpha_{hl} [f_{BE}(\mu_{hl}) - f_{pt}(\infty)] E^2 dE dVol / A \}
\end{aligned}
$$

$$\tag{8.6}$$

The first two terms are modified since only a fraction of the incoming photons are absorbed by the targeted process. A similar modification occurs in the third term since only a fraction of the emitted photons at any given energy are supplied by this process. Total photon emission at any energy, however, is determined by the equilibrium photon distribution function (Figs. 4.7 and 4.8), for a reasonably thick device. The fourth term is new and is due to the breakdown of photon recycling. The integral will be positive for processes with high chemical potentials but negative for those where this parameter is low.

The first component of this fourth term represents the generalised Shockley-Van Roosbroek equation (Eq. (4.8)) for radiative recombination by the chosen process at any given point in the cell volume. The second term relates to the fraction of the total number of photons of a given energy produced internally at a given point that are captured by the targeted process.

Since the fourth term makes a negative contribution to device current for high chemical potential processes and a positive contribution for low chemical potentials (so that the net sum to all equals zero), it follows that the term represents a net energy loss as expected. It is volume dependent as previously noted for other loss terms, such as those due to finite mobility or non-radiative recombination or free carrier absorption.

8.5 Other Possible Multigap Schemes

Several ideas for implementing multigap cells have already been discussed (Fig. 8.7). Are there other ideas that might allow extension of the approach to devices with more than three bands?

One possibility is to chose existing semiconductor material where the band structure consists of several narrow bands. For example, the elemental semi-conductors Se and Te are thought to have such a band structure, due to the splitting of 3 p-levels that are degenerate (i.e., have the same energy) in isolated atoms of the material. Various binary semiconductors including I-VII and I_3-VI compounds are also thought to have similar properties (Berger 1997).

One difficulty with these being semiconductors is that excitations between fully empty or between two full bands are likely to be weak. By degenerately doping such material, this limitation could be overcome, suggesting up to 4-band operation is feasible with this approach (Fig. 8.8).

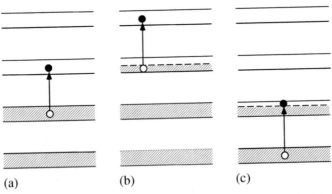

(a)   (b)   (c)

Fig. 8.8: Possible 4-band operation of semiconductor material with multiple narrow conduction and valence bands using degenerate doping.

Another approach may be to use low dimensional structures. For example, Fig. 8.9 shows a semiconductor superlattice where minibands formed in the conduction band are used to form a 3-band cell. Actually in the superlattice shown, energy confinement occurs only in the direction perpendicular to the superlattice. In the direction into the page, carriers can make lateral phonon-assisted transitions between the bands, without emitting a photon. This would detract from the desired operation of the device unless such transition rates can be kept small compared to radiative rates. This is not impossible, as attempts to make lasers based on transitions between such minibands confirm.

Alternatively, a three dimensional layout of closely spaced quantum dots would ensure the desired mode of operation. To produce such structures is currently technologically quite challenging, although much work is being conducted seeking the self-assembly of such structures. Note that, earlier, a regular 3-D arrangement of impurity atoms was mentioned as likely to give the best impurity band properties, showing some commonality in requirements between seemingly different approaches.

The previous suggestions give some confidence that a practical approach can be found to extending the 3-band approach to more than three bands. There are very definite benefits in doing so, as explored in the next section.

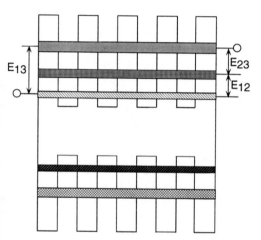

Fig. 8.9: Proposed 3-band cell based on superlattice minibands. A superlattice of quantum dots is ideally required.

The analysis of Sects. 8.2 and 8.4 can be simply extended to the 4-band cell of Fig. 8.10(a). For example, Eq. (8.1) will still hold. The main modifications will be to the requirements to match currents and voltages.

The relations for current continuity in the steady state are:

$$I_{21} = I_{32} + I_{42} \tag{8.7}$$

$$I_{43} = I_{31} + I_{32} \tag{8.8}$$

while the chemical potentials are related by:

$$\mu_{41} = \mu_{21} + \mu_{32} + \mu_{42} = \mu_{21} + \mu_{42} = \mu_{43} + \mu_{31} \tag{8.9}$$

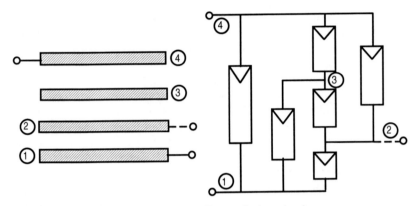

Fig. 8.10: (a) 4-band cell; (b) corresponding equivalent circuit.

These relations can best be interpreted in terms of the equivalent circuit of Fig. 8.10(b). Maximising output involves a multi-variable optimisation of the energy separations between and widths of the bands. However, the interconnection apparent in the equivalent circuit provides sufficient flexibility for all cells to span the solar spectrum, convert their reasonable share of photons and to operate at close to their optimal operating point. The total power delivered is therefore not be appreciably lower than in a tandem cell design involving a comparable number of individually contacted cells.

Note, however, that, as the number of bands, N, increases, the total number of cells increases as $N (N - 1)/2$. Hence, the 4-band case approaches the performance of a 6 cell tandem, with the problem of contacting them already solved and the redundancy in connections providing a much better tolerance to spectral change than if series connected (Brown and Green 2002b; Brown et al. 2002).

8.6 Impurity Photovoltaic Effect

The previous theory can be extended to the impurity photovoltaic effect of Fig. 8.1 by noting a theorem proposed elsewhere (Green 2001). Entropy generation is minimised if the level of excitation within the absorber (as measured by the chemical potential associated with the inverse spontaneous emission process) is constant over the absorption volume.

The general proof of this proposition is quite unwieldy (Green 2002) although it is simpler to prove in specific examples (e.g., Exercise 8.2).

For the case of a single impurity level, the consequence of this theorem is that the occupancy of each separate defect site must be equal for limiting performance under the general set of assumptions used throughout this book. This means that the chemical potential associated with excitations between the defects and the conduction band, for example, will have the same value throughout the device.

Invoking this theorem therefore greatly simplifies the analysis. The results are that, for a single defect species, the limiting efficiency is 63.2%, identical to the impurity band case (Brown and Green 2000a). If a large number of different types of defects are considered, each type going a separate energy level, the limiting efficiency increases to 77.2%

8.7 Up-and Down-Conversion

A disadvantage of introducing impurities in the bulk of the cell is that they are likely, in practice, to introduce additional non-radiative recombination paths (Guttler and Queisser 1970). However, if the associated optical excitations are decoupled from the cell, this objection can be overcome (Trupke et al. 2002a; 2002b). Surprisingly, a down-converter can be placed on the front surface of a cell, creating two photons from high energy photons without significantly interfering with the transmission of lower energy photons into the cell (Trupke et al. 2002a). An up-converter placed on the rear of the cell can create one high energy photon by a multi-step excitation involving two lower energy photons. Surprisingly, the efficiency of this process is enhanced if there is energy relaxation during the process (Trupke et al. 2002b).

Exercises

8.1   Consider the 3-band cell design of Fig. 8.3. For the case of unconcentrated solar radiation, and impurity bandwidth approaching zero, calculate (assuming $T_s = 6000$ K, $T_c = 300$ K):
(a)   The maximum steady-state current per unit illuminated area able to be excited between each pair of bands for a positive chemical potential difference between them (this corresponds to the short-circuit current densities of the "cells" in the equivalent circuit of Fig. 8.4(a)).
(b)   The variation of this current with varying positive chemical potential assuming the Shockley-Queisser approximation. Comment on the validity of this approximation.
(c)   Graphically or otherwise, find the variation of steady-state current per unit illuminated area that can be excited through the impurity band as a function of the chemical potential difference between the conduction and valence band. (Graphically, this can be done by noting the equivalent circuit of Fig. 8.4(a), plotting the current-chemical potential curves of the two devices in series modeling excitations through the impurity band and then finding their composite curve, by adding potential for each fixed value of current).
(d)   Graphically or otherwise, find the variation of total steady-state current per unit illuminated area that can be excited between the conduction and

valence band. (Graphically, the current-chemical potential curve corresponding to direct excitations can be plotted on the same graph as resulted from part (c), above, and the composite curve found by adding currents at each fixed value of potential).

(e)   Using the above results or otherwise, find the energy conversion efficiency at the maximum power point for this design.

8.2   (Difficult). Consider the limiting performance of the "same bandgap" tandem cell as sometimes used, for practical reasons, in amorphous silicon cell technology. A cell of finite thickness is placed on top of a thicker cell of the same bandgap (assume infinitely thick and that the cells are separated by an air gap and that they are not series-connected).

For idealised cells in the radiative limit, is it possible to gain a performance advantage using this configuration, (e.g., by operating the cells at different voltages)?

References

Araujo GL and Marti A (1995), Electroluminescence coupling in multiple quantum well diodes and solar cells, Appl Phys Lett 66: 894-895.

Barnham K and Duggan G (1990), A new approach to high efficiency multi-bandgap solar cells, J Appl Phys 67: 3490-3493.

Berger LI (1997), "Semiconductor Materials", CRC Press, Boca Raton, Florida.

Brown AS and Green MA (2002a), Limiting efficiency for current-constrained two-terminal tandem cell stacks, Progress in Photovoltaics 10: 299-307.

Brown AS and Green MA (2002b), Impurity photovoltaic effect: fundamental energy conversion efficiency limit, Journal of Applied Physics 92: 1329-1336.

Brown AS, Green MA and Corkish R (2002), Limiting efficiency for multi-band solar cells containing three and four bands", Physica E 14: 121-125, April.

Capasso F et al. (1985), Appl Phys Lett 47: 420.

Corkish R (1999), Calculations conducted for the author.

Cuadra L, Marti A and Luque A (2000), Modelling of the absorption coefficient of the intermediate band solar cell, paper presented at 16th European Photovoltaic Solar Energy Conference, Glasgow, May, 15-21.

Graetzel M (2000), Perspectives for dye-sensitized nanocrystalline solar cells, Progress in Photovoltaics 8: 171-186.

Green MA (1999), "Third Generation Photovoltaics", Grant Application prepared for Australian Research Council's Special Research Centres Scheme, February.

Green MA (2000), Journal Material Science B.

Green MA (2002), Conditions for minimum entropy production during light absorption and application to advanced solar conversion concepts, QUANTSOL 2002, Rauris, Austria, March.

Guttler G and Queisser HJ (1970), Impurity photovoltaic effect in silicon, Energy Conversion 10: 51-55.

Keevers MJ and Green MA (1994), Efficiency improvements of silicon solar cells by the impurity photovoltaic effect, Journal of Applied Physics 75: 4022-4033.

Luque A, Marti A. and Cuadra L (2000), High efficiency solar cell with metallic intermediate band, paper presented at 16th European Photovoltaic Solar Energy Conference, Glasgow, May, 59-62.

Luque A and Marti A (1997), Increasing the efficiency of ideal solar cells by photon induced transitions at intermediate levels, Physical Review Letters 78: 5014-5017.

Luque A, Marti A Cuadra L (2001), thermodynamic consistency of sub-bandgap absorbing solar cell, IEEE Trans. ED-48: 2118-2124, September.

Milne AG (1973), "Deep impurities in semiconductors", Wiley, New York.

Shklovskii BI and Efros AL (1960), "Electronic Properties of Doped Semi-conductors", Springer-Verlag, Berlin.

Trupke T, Green MA and Würfel P (2002a), Improving solar cell efficiencies by down-conversion of high-energy photons, J Appl Phys, 92: 1668-1674.

Trupke T, Green MA and Würfel P (2002b), Improving solar cell efficiencies by the up-conversion of sub-band-gap light, J Appl Phys 92: 4117-4122.

Wolf M (1960), Limitations and possibilities for improvement of photovoltaic solar energy converters, Proceedings of the IRE 48: 1246-1263.

Wurfel P (1993), Limiting efficiency for solar cells with defects from a three-level model, Solar Energy Materials and Solar Cells 29: 403-413.

9 Thermophotovoltaic and Thermophotonic Conversion

9.1 Introduction

One way of reducing the energy loss on light absorption that occurs in a conventional cell is to reduce the average energy of the photon absorbed by it. This is the basic idea behind thermophotovoltaic conversion, schematically shown in Fig. 9.1. Sunlight (or heat from another source) is absorbed in a receiver that is thereby heated to a reasonably high temperature, generally much lower than that of the sun. The heated receiver then radiates energy to a photovoltaic cell. Although many of the photons emitted by the receiver may be too low in energy to be used by the cell, these can be reflected by the cell back to the receiver, helping to maintain its temperature. As a result, these photons are not necessarily wasted. Light emitted by the cell is also not wasted. The absorption and emission properties of the receiver also can be manipulated to improve performance having them different on either side. Different signs of the absorption and emission areas are also possible.

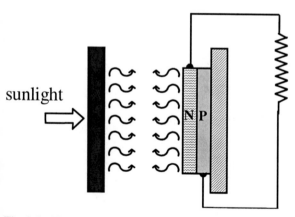

Fig. 9.1: Thermophotovoltaic conversion.

Initially, thermophotovoltaics was suggested as a way of converting heat to electricity (Fortini 1962; Wedlock 1963). A focussed developmental program in the late 1970s, however, attempted to demonstrate high efficiency solar conversion with this approach (Swanson 1979). Although not leading to a practical system, this program resulted in the development of silicon concentrator solar cells based on new design principles, that still hold the record for direct solar conversion efficiency (Swanson et al. 1984). Interest in the approach has recently re-emerged in the context of developing compact power sources for spacecraft, using fossil fuels as the primary energy source (Coutts and Fitzgerald 1998). The simple geometry of Fig. 9.1 will be used as a the starting point for our analysis, although optimal geometries can be quite complex, particularly for direct sunlight conversion, as reviewed elsewhere (Davies and Luque 1994).

A recent suggestion is that of thermophotonic conversion shown in Fig. 9.2 (Green and Wenham 1998; Harder et al. 2003).   This approach arose from an attempt to use the refrigerating action of an ideal light emitting diode (LED) to improve photovoltaic performance. Such a diode ideally can emit a photon of energy above the LED's bandgap by the supply of an electron at a potential only a fraction of the bandgap potential. This seeming paradox is resolved by the fact that the diode absorbs heat from its junction and contacts to supply the energy balance (Dousmanis et al. 1964; van der Ziel 1976). The configuration shown in Fig. 9.2 is almost symmetrical, except that heat is supplied to one diode to maintain it at a higher temperature than the other. The latter is maintained at ambient temperature with the two diodes optically coupled, but thermally isolated. When both diodes operate at the radiative limit, heat supplied to the hot device ideally is converted to power in an electrical load connected between them. The efficiency of this transformation can approach the Carnot limit for conversion between the temperatures of the hot and cold device.

As a preliminary to investigating the limiting performance of these approaches, the performance of absorbers in sunlight first will be examined.

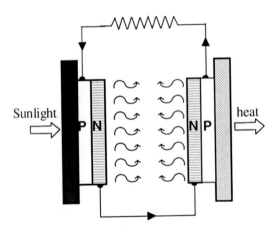

Fig. 9.2: Thermophotonic conversion (conceptual only).

9.2 Solar Thermal Conversion

To get a feel for desirable absorber properties, the problem treated in Exercise 2.2 will be examined in more detail. This problem involved the system of Fig. 9.3, where the maximum solar conversion efficiency was calculated for a system based on a receiver with good absorption properties on one side, and low emission properties on the other.

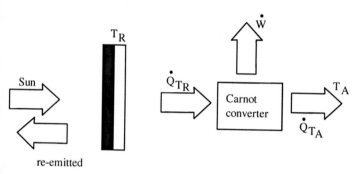

Fig. 9.3: Photothermal system based on the conversion of heat collected by an absorber at the Carnot efficiency.

For direct sunlight conversion ($f_s = f_c$), the maximum possible conversion efficiency for black-body absorption properties is given by the familiar result:

$$\eta = [1-(T_R/T_S)^4](1-T_A/T_R) \qquad (9.1)$$

The first term represents the net radiative heat input to the receiver (that from the sun minus that re-emitted), while the second term represents the conversion of this heat to work at the Carnot limit for a system operating between the receiver and ambient temperatures (T_R and T_A, respectively). Although this equation predicts an encouraging efficiency of 85.4% for direct sunlight for the absorber temperature of 2544 K, the corresponding calculation for diffuse sunlight gives a more modest efficiency of 11.7%.

How can the latter poor performance be overcome? A well-known technique in the solar thermal area is the use of energy selective absorbers. These have good absorption properties at the high energies typical of the peak emission from sunlight, but poor absorption and hence emission properties at the lower energies, where a non-selective absorber would have peak emission. This reduces absorber emission while still maintaining reasonable solar input. The optimum design is to have a sharp transition between zero and unity absorption at a certain energy threshold, E_t. The energy balance in this case becomes:

$$\dot{Q} = f_s\dot{E}(E_t,\infty,0,T_s)+(f_c-f_s)\dot{E}(E_t,\infty,0,T_A)-f_c\dot{E}(E_t,\infty,0,T_R) \qquad (9.2)$$

where $\dot{Q}$ is the heat flow into the Carnot converter. The overall efficiency as a function of E_t turns out to be identical to that given by the earlier analysis (Würfel 1997), corresponding to a hot carrier cell with zero chemical potential (Sect. 6.5). The results are shown in Fig. 6.7 for the case where $\Delta\mu_H = 0$.

While energy selective absorbers do not significantly improve prospects for the photothermal conversion of direct sunlight, they do have a big impact on diffuse sunlight conversion efficiency, increasing limiting efficiency to 53.6% for a threshold energy of 0.90 eV (de Vos 1992).

Several techniques can be used to produce selective absorbers, including some that are well known in the photovoltaics field, such as the use of interference coatings (Höfler et al. 1983) and surface texturing (Davies and Luque 1994). Semiconductor material itself makes a good selective absorber. Some of the best performing selective absorbers have been based in SiGe alloys (Mills 2000), capable of threshold energies near the diffuse conversion peak of Fig. 6.7.

9.3 Thermophotovoltaic Conversion

9.3.1 Black-Body Source

If the receiver is assumed to have ideal black-body properties on both sides, heat will be extracted from it by net radiative transmission to the cell. The analysis of this case obviously is closely related to that for solar energy conversion where the sun is the black-body radiator, except photons can be recycled. Assuming the ideal case where the cell reflects all photons of energy below its bandgap but absorbs all photons of energy above it, a particle balance leads to the following expression for power output per unit area:

$$P = qV[\ \dot{N}(\ E_G,\infty,0,T_R\ ) - \dot{N}(\ E_G,\infty,qV,T_c\ )]\qquad (9.3)$$

The net rate of heat supply from the receiver becomes:

$$\dot{Q} = [\ \dot{E}(\ E_G,\infty,0,T_R\ ) - \dot{E}(\ E_G,\infty,qV,T_c\ )]\qquad (9.4)$$

Hence the conversion efficiency of heat from the absorber becomes:

$$\eta_Q = qV\ \frac{\dot{N}(\ E_G,\infty,0,T_R\ )}{\dot{E}(\ E_G,\infty,0,T_R\ )}\ \frac{[1 - \dot{N}(\ E_G,\infty,qV,T_c\ )/\dot{N}(\ E_G,\infty,0,T_R\ )]}{[1 - \dot{E}(\ E_G,\infty,0,T_c\ )/\dot{E}(\ E_G,\infty,0,T_R\ )]}\qquad (9.5)$$

where, from Appendix D,

$$\frac{\dot{E}(E_G,\infty,0,T_R)}{\dot{N}(E_G,\infty,0,T_R)}$$

$$= E_G \frac{[1+3(kT_R/E_G)\beta_1/\beta_0+6(kT_R/E_G)^2\beta_2/\beta_0+6(kT_R/E_G)^3\beta_3/\beta_0]}{[1+2(kT_R/E_G)\beta_1/\beta_0+2(kT_R/E_G)^2\beta_2/\beta_0]}$$

(9.6)

The argument of all the β functions is $(-E_G/kT_R)$. The above shows that the average energy of the absorbed photon is close to $E_G + kT_R$, when $kT_R << E_G$. Also:

$$\frac{\dot{N}(E_G,\infty,qV,T_c)}{\dot{N}(E_G,\infty,0,T_R)} = \frac{T_c\beta_0^*[1+2(kTc/E_G)\beta_1^*/\beta_0^*+2(kT_c/E_G)^2\beta_2^*/\beta_0^*]}{T_R\beta_0[1+2(kT_R/E_G)\beta_1/\beta_0+2(kT_R/E_G)^2\beta_2/\beta_0]}$$

(9.7)

where the argument of the β^* function is $-(E_G-qV)/kT_c$, and:

$$\frac{\dot{E}(E_G,\infty,qV,T_c)}{\dot{E}(E_G,\infty,0,T_R)} =$$

$$\frac{T_c}{T_R}\frac{\beta_0^*}{\beta_0}\frac{[1+3(kT_c/E_G)\beta_1^*/\beta_0^*+6(kT_c/E_G)^2\beta_2^*/\beta_0^*+6(kT_c/E_G)^3\beta_3^*/\beta_0^*]}{[1+3(kT_R/E_G)\beta_1/\beta_0+6(kT_R/E_G)^2\beta_2/\beta_0+6(kT_R/E_G)^3\beta_3/\beta_0]}$$

(9.8)

Note that the ratios given in Eqs. (9.7) and (9.8) are very close to being equal, with that given by Eq. (9.7), slightly larger by a factor of order $(1 + kT_R/E_G)$, again when $kT_R << E_G$. Under these conditions:

$$\eta_Q \approx \frac{qV[1-(kT_c/E_G)\beta_0^*/\beta_0]}{E_G(1+kT_R/E_G)}$$

(9.9)

where

$$\frac{\beta_0^*}{\beta_0} = \frac{ln\{1-exp[-(E_G-qV)/kT_c]\}}{ln\{1-exp[-E_G/kT_R]\}}$$

(9.10)

For $(E_G - qV)>>kT_c$, Eq. (9.10) reduces to the ratio of the exponentials involved and Eq. (9.9) reverts to a form similar to the standard ideal solar cell equation. However, unlike standard solar conversion, maximum efficiency values are predicted as the bandgap approaches infinity, where the efficiency approaches the Carnot limit.

This difference from when the sun is the illuminating black-body source is due mainly to the assumed efficiency in photon recycling in the present case. Wide bandgap material will be very sensitive to the efficiency of such recycling for the huge quantities of sub-bandgap photons involved. Since these are not used, reducing their numbers will improve the practicality of the proposed process. For example, if the emission surface of the receiver were made from material with the same bandgap as the cell, the emission spectrum would contain a reduced number of these photons.

This leads to the concept of energy selective emitters (Coutts and Fitzgerald 1998) that emit light only over narrow bandwidth ranges. One of the most important are rare-earth doped ceramics (Krishna et al. 1999). The strong radiative transitions, that make rare-earth doped crystals of interest in lasers, also are activated in these selective emitters, concentrating the emitted energy over a narrow bandwidth range.

The corresponding effect of selective emission on performance in the limit can be analysed using the configuration of Fig. 8.5, where a narrow passband filter is placed between the emitter and cell. This filter is assumed only to pass light (in both directions, by reciprocity) in a narrow energy range ΔE, beginning at the energy E_f. All other energies are assumed to be perfectly reflected.

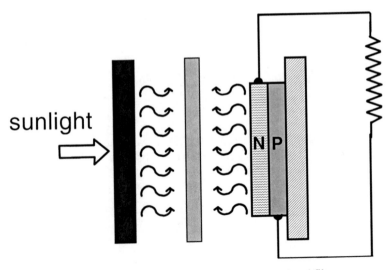

Fig. 9.4: Thermophotovoltaic conversion with narrow passband filter.

9.3.2 With Narrow Passband Filter

The analysis, with the filter included, parallels that in the previous section, except that only the integrands, rather than full integrals, are involved. The ratios appearing in Eq. (9.5) become:

$$\frac{\dot{E}(E_f, E_f', 0, T_R)}{\dot{N}(E_f, E_f', 0, T_R)} = E_f \qquad (9.11)$$

$$\frac{\dot{N}(E_f,E'_f,qV,T_C)}{\dot{N}(E_f,E'_f,0,T_R)} = \frac{\dot{E}(E_t,E'_f,qV,T_C)}{\dot{E}(E_f,E'_f,0,T_R)} \tag{9.12}$$

where $E'_f = E_f + \Delta E$. Hence, the efficiency is merely:

$$\eta_Q = qV / E_f \tag{9.13}$$

where qV can be as large as $E_f(1 - T_C/T_R)$, leading to a Carnot efficiency in principle regardless of the cells bandgap when energy transfer approaches zero. Ideally, E_f would be chosen to be just at the semiconductor bandgap E_G.

9.3.3 Solar Conversion: Cell/Receiver

Since the cell/receiver combination can be designed to convert at close to the Carnot efficiency, it might be thought that illuminating the receiver with sunlight would lead immediately to efficiencies equal to the photothermal conversion limits of Sect. 9.2. However, an additional requirement is to match the optimal rate of supply of energy from the sun to the receiver with the optimal rate from the receiver to the cell. The cell operates at close to the Carnot limit only when the energy reaching it is suppressed, using either large band-gaps or selective sources or filters.

This mismatch can be accommodated by using different areas for the absorber, emitter and cell. Some simple extensions of the flat absorber treated so far are shown in Fig. 9.5. When only direct light is of interest, a range of more sophisticated optical approaches is possible, as reviewed elsewhere (Davies and Luque 1994).

With this flexibility, thermophotovoltaics can, in principle, obtain the same energy conversion efficiencies as solar photothermal conversion. The results of Fig. 6.7 still apply, with the threshold that of the absorber rather than that of the

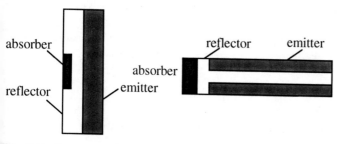

Fig. 9.5: Some examples of simple designs with different absorption and emission areas. Cell area approximately equals emission area in these designs. Design (b), with minor modifications would be suitable for diffuse light conversion.

filter. Properties of the cell, emitter and any filters can be selected independently of this threshold, provided heat transfer between the absorber and emitter is not entirely by photons (e.g, some by electrons or phonons).

9.4 Thermophotonics

9.4.1 Case with Filters

The simplest case to analyse is that with filters and independent voltage sources to bias the LED and solar cell as shown in Fig. 9.6.

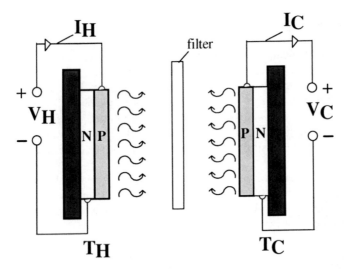

Fig. 9.6: Thermophotonic conversion, with narrow pass filter and independent power supplies.

The power supply to the LED boosts the light emission compared to thermal emission. Since V_H will, for an ideal device, be less than E_G/q, the photons emitted by this device will have more energy than that drawn from this supply, with the difference drawn from the heat source. An energy balance gives:

$$\dot{Q}_H + V_H I_H / A = \dot{E}_H - \dot{E}_C \tag{9.14}$$

where $\dot{E}_H$ and $\dot{E}_C$ are the photon energy fluxes emitted and A is the device area. In the radiative limit, a particle balances gives:

$$I_H / (qA) = \dot{N}_H - \dot{N}_C \tag{9.15}$$

where $\dot{N}_H$ and $\dot{N}_C$ are the photon fluxes emitted by the two devices. Applying the same balance to the cell shows that, in the radiative limit:

$$I_C = I_H = I_L \tag{9.16}$$

If recombination processes additional to radiative combination are present, then the current supplied to the LED will be higher than that given by the difference in the photon fluxes, while that generated by the cell will be lower. However, in the radiative limit, the net electrical power out is given by:

$$P = V_C I_C - V_H I_H = (V_C - V_H)\, I_L \tag{9.17}$$

Hence, the efficiency of conversion to electricity of the heat supplied becomes:

$$\eta = \frac{P/A}{\dot{Q}_H} = \frac{q(V_C - V_H)(\dot{N}_H - \dot{N}_C)}{[\dot{E}_H - \dot{E}_C - qV_H(\dot{N}_H - \dot{N}_C)]} \tag{9.18}$$

When the filter is in place, the expressions for the energy and particle fluxes simplify. In particular:

$$(\dot{E}_H / \dot{N}_H) = (\dot{E}_L / \dot{N}_L) = E_f \tag{9.19}$$

and

$$V = E_f - \frac{kT}{q} \ln\left[\frac{2\pi E_f^2 \Delta E}{h^3 c^2 \dot{N}}\right] \tag{9.20}$$

The efficiency becomes:

$$\eta = \frac{V_C - V_H}{E_f / q - V_H} = 1 - \frac{T_C}{T_H}\frac{\ln(1 + \dot{N}_L / \dot{N}_C)}{\ln(1 + \dot{N}_L / \dot{N}_H)} \tag{9.21}$$

where $\dot{N}_L = [2\pi E_f^2 \Delta E /(h^3 c^2)]$. The efficiency conversion approaches the Carnot limit when the current in the load equals a small imbalance between two large photon emission currents. The heat input converted equals:

$$\dot{Q}_H = kT_H \ln(1 + \dot{N}_L / \dot{N}_H)(\dot{N}_H - \dot{N}_C) \tag{9.22}$$

For each value of V_C, a value for V_H can be found, in principle, that maintains the hot device at the desired operating temperature T_H provided the heat supply rate is between $-E_f \dot{N}_C$ up to $kT_H \dot{N}_L$. The optimum value of V_C will be as high as practical, giving:

$$\dot{N}_C / \dot{N}_H = 1 - \dot{Q}_H /(kT_H \dot{N}_L) \tag{9.23}$$

The limiting efficiency becomes:

$$\eta = 1 - \frac{T_C}{T_H} / (1 - \frac{\dot{Q}_H}{kT_H \dot{N}_L}) = 1 - \frac{T_C}{T_H} / (1 - \frac{\dot{Q}_H h^3 c^2}{2\pi k T_H E_f^2 \Delta E}) \qquad (9.24)$$

Note the conflicting requirements on ΔE. It has to be small for the present analysis to apply, but large to allow high values of $\dot{N}_L$.

Having found the operating voltages and currents, the negative terminals of the batteries shown could be connected and a resistor of value $(V_C - V_H)/I_L$ used to connect their positive terminals. The same operating conditions would be maintained except that, in the radiative limit, no current would be drawn from the batteries. These could be removed, leading to the connection of Fig. 9.2.

If the two devices in this system start with both at room temperature, and heat is supplied to the LED at a predetermined rate, what determines the final temperature reached by the LED? This final condition is not uniquely defined by the value of the resistor in the circuit, since a range of possible operating temperatures are possible for each combination of heat supply rate and R. The stationary state reached by the system will be that of minimum entropy production, consistent with the external constraints on the system (de Groot and Mayer 1984).

Taking a global perspective on the system of Fig. 9.2, heat at rate $\dot{Q}_H$ is supplied as an input at temperature T_H. Outputs are electrical work in the resistor in the resistor and heat rejected at temperature, T_C with the corresponding rates equal to $\dot{W}$ and $(\dot{Q}_H - \dot{W})$. The overall rate of entropy production is therefore:

$$\dot{S} = \frac{\dot{Q}_H - \dot{W}}{T_C} - \frac{\dot{Q}_H}{T_H} \qquad (9.25)$$

This will always be positive or zero, becoming only zero only for systems operating at the Carnot limit. Substituting expressions for the conversion efficiency $\dot{W}/\dot{Q}_H$ will allow the entropy generation rate to be found for the specific examples studied. For example, substituting Eq. (9.24) into Eq. (9.25) gives:

$$\dot{S} = \frac{\dot{Q}^2 H}{kT_H^2 \dot{N}L} N (1 - \frac{\dot{Q}_H}{kT_H \dot{N}_L}) \qquad (9.26)$$

The minimum value occurs only at the highest possible T_H consistent with the energy transfers and other constraints involved. This mitigates against the chemical potential enhanced emission modes of operation previously explored.

If the all recombination in both devices does not occur radiatively, the equality of the LED and cell currents is no longer maintained. The cell with independent power supplies would operate with reduced efficiency. However, the circuit of Fig. 9.2 could not operate in the bias region previously explored. The LED would require more current to reach this region than the cell is capable of supplying.

This situation is modelled in Fig. 9.7. To maintain particle and current continuity, the current in the parasitic diodes modelling non-radiative recombination must be equal and opposite, forcing one device into reverse bias. The LED would switch to a higher temperature, negative chemical potential mode of operation where the photon flux emission was suppressed below that of a black-body at the same temperature. The non-radiative processes would result in additional carrier generation in this mode that would add to the current supplied from the cell. The cell would operate at a positive chemical potential, with its non-radiative current loss balanced by the generation current in the LED. This is consistent with the conclusion previously reached for the lossless case. From the above, it is concluded that active bias or temperature control is essential to reach the desired high positive chemical potential mode of operation for the light emitting device.

The reversed bias mode of operation of the LED offers the potential of conversion at close to the Carnot efficiency and of other features such as power delivered at a higher potential than the semiconductor bandgap. However, the suppressed emission from the LED means lower possible heat conversion rates for any given operating temperature. Free carrier absorption will determine the emittance of an actual device in this mode, destroying the selective energy emission features of the LED that is such an attractive feature of thermophotonics (Lin et al. 2002).

The configuration of Fig. 9.6 using chemical batteries as the voltage source would be one possible option. If one polarity terminal of the batteries were connected together, the electrical load could be placed between the other two, with its value determining the rate of charge or discharge of the two batteries involved. Alternatively, it may be possible to replace the batteries by capacitors with a suitably controlled load. Another option would be to connect the load across the cell, as in traditional systems, with a controller equalising the battery or capacitor charge.

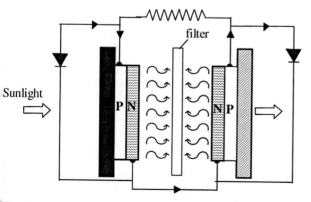

Fig. 9.7: Thermophotonic conversion with non-radiative recombination modelled by parasitic non-emitting diodes.

9.4.2 Without Filter

If the filter is removed, the analysis parallels that of the previous section with Eq. (9.18) remaining valid. However, the simplification made possible by Eqs. (9.19) and (9.20) no longer apply. Instead, the equivalent of Eq. (9.21) becomes:

$$\eta = \frac{V_C - (1 - \dot{N}_C / \dot{N}_H)V_H}{(1 - \dot{E}_C / \dot{E}_H)\dot{E}_H /(q\dot{N}_H) - (1 - \dot{N}_C / \dot{N}_H)V_H} \tag{9.27}$$

Where $\dot{E}_H / \dot{N}_H$, $\dot{N}_C / \dot{N}_H$ and $\dot{E}_C / \dot{E}_H$ are given by the same expressions as Eqs. (9.6), (9.7) and (9.8) but where T_R is replaced by T_H and the argument of the β-functions equals $- (E_G - qV_H/kT_H)$, while that of the β^* functions equals $- (E_G - qV_C)/kT_C$. In the most efficient mode of operation, these arguments approach zero, with the ratios approaching, for $kT_H << E_G$:

$$\dot{E}_H / \dot{N}_H \approx E_G[1 + (kT_H / E_G)\beta_1 / \beta_0] \tag{9.28}$$

$$\dot{N}_C / \dot{N}_H \approx (T_C / T_H)(\beta_0^* / \beta_0)[1 + 2(kT_C / E_G)\beta_1^* / \beta_0^* - 2(kT_H / E_G)\beta_1 / \beta_0] \tag{9.29}$$

$$\dot{E}_C / \dot{E}_H = (T_C / T_H)(\beta_0^* / \beta_0)[1 + 3(kT_C / E_G)\beta_1 / \beta_0^* - 3(kT_H / E_G)\beta_1 / \beta_0] \tag{9.30}$$

Again, efficiency close to the Carnot limit is feasible. The main loss is due to the thermalisation of carriers emitted at temperature T_H to the temperature T_C on absorption in the cell, although this loss applies only to the net carrier flow.

Exercises

9.1   For representative parameters of bandgaps and operating temperatures, calculate limiting efficiencies and compare to the Carnot limit for the situation analysed in Sect. 9.4.2.

References

Coutts T and Fitzgerald M (1998), Thermophotovoltaics, Scientific American, September.

Davies PD and Luque A (1994), Solar thermophotovoltaics: brief review and a new look, Solar Energy Materials and Solar Cells 33: 11-22.

de Groot SR and Mazur P (1984), Non-Equilibrium Thermodynamics, Dover, New York.

de Vos A (1992), Endoreversible Thermodynamics of Solar Energy Conversion, Oxford University Press.

Dousmanis GC Mueller CW, Nelson H and Petzinger KG (1964), Evidence of refrigerating action by means of photon emission in semiconductor diodes, Physical Review 133: A316-318.

Green MA and Wenham SR (1998), unpublished work.

Harder, N-P and Green MA (2003), Thermophotonics, invited paper to Special Edition of Semiconduct Sc & Techn (to be published).

Höfler H, Paul HJ, Würfel P and Ruppel W (1983a), Selective emitters for thermophotovoltaic solar energy conversion, Solar Cells 10: 257-271.

Höfler H, Paul HJ, Ruppel W and Würfel P (1983b), Interference filters for thermophotovoltaic solar energy conversion, Solar Cells 10: 273-286.

Krishna MG, Rajendran M, Pyke DR and Bhattacharya AK (1999), Spectral emissivity of Ytterbium oxide-based materials for application as selective emitters in thermophotovoltaic devices, Solar Energy Materials and Solar Cells 59: 337-348.

Lin KL, Catchpole KR, Trupke T, Green MA, Aberle AG and Corkish R (2002), Thin semiconducting layers as selective emitters in thermophotonics systems, Conference Proceedings, 29th IEEE Photovoltaics Specialists Conference, New Orleans, to be published.

Mills D (2000), Private communication, Sydney University.

Swanson RM (1979), A proposed thermophotovoltaic solar energy conversion system, Proc IEEE 67: 446-447.

Swanson RM, Beckwith SK, Crane RA, Eades WD, Kward YK, Sinton RA and Swirhun SE (1984), Point contact silicon solar cell, IEEE Trans. Electron Devices ED-31: 661-664.

van der Ziel A (1976), Solid State Physical Electronics, 3rd ed., Prentice-Hall, New Jersey, 425.

Wedlock BD (1963), Thermophotovoltaic energy conversion, Proc IEEE 51, 694-698; also A Fortini (1962), "La Conversion D'Energie Thermo-photo-electrique, L'Onde Electrique 442: 530-540.

Würfel P (1997), Solar energy conversion with hot electrons from impact ionisation, Solar Energy Materials and Solar Cells 46: 43-52.

10 Conclusions

From the analysis of this book, it appears that there are sufficient options for improving the performance of solar photovoltaic cells beyond the single junction limits, that greatly improved performance, at some stage in the future, is very likely.

The tandem cell approach of Chap. 5 already demonstrates that such enhanced performance is feasible. Cell technologies right at the top in terms of single junction cell performance and right at the bottom have already benefited from the tandem approach. Experimental gains to date have been in the 20-25% range, relative to single junction devices, compared to theoretically achievable boosts of over 100%. For tandems involving a large number of cells, a generic approach to tandem cell design would be desirable. An example is shown in Fig. 10.1, where bandgaps within a single materials system are controlled by varying superlattice spacings is an example of such an approach (Fig. 10.1).

The hot carrier cells of Chap. 6 offer performance potential very similar to the case of an infinite number of tandem cells, but may have major operational advantages, as well as not requiring as much effort to fabricate. One major challenge is to develop suitable technology for forming energy selective contacts. Energy selective tunnelling, such as in resonant tunnelling devices, might provide a promising approach here. Narrow conductive bands within a material with a large forbidden gap, as devised for some of the other conversion options, would be another option as would conduction to valence band tunneling used in tunnel diodes (Conibeer et al 2002). The second major challenge is to find a way of maintaining hot carrier concentrators by accelerating radiative processes relative to energy relaxation.

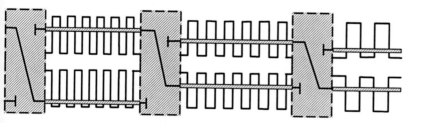

Fig. 10.1: Generic tandem cell design based on superlattices.

Without an as yet unidentified conceptual breakthrough, performance enhancement by creating multiple electron hole pairs per photon (Chap. 7) does not seem promising. All suggested processes so far are too weak and inefficient to offer much scope. Some new concepts are required here.

The multiple energy threshold approaches of Chap. 8 would seek to offer exciting prospects due to likely improvements in the materials engineering area. Three dimensional control of material structure seems particularly relevant to concepts such as the multiband approach. Such control is likely to improve dramatically over the coming two decades. Photon up- and down-conversion without electrical contact to the elements performing this function also appears to have considerable potential.

Thermophotovoltaics (Chap. 9) is already a very active area of research. Improved energy selectivity in emission appears to be a key requirement here. The main advantage of thermophotonics appears to be larger feasible energy transfers at relative low operating temperatures. This may make the approach very well suited for the conversion of low grade heat at efficiencies close to the Carnot limit. However, considerable improvement in the radiative efficiency of both inexpensive light emitters and solar cells is required before this could become a practical option.

With the accelerating pace of materials science development, many of the ideas outlined in the preceding chapters that now appear highly speculative are likely to become implementable experimentally. By exploring desirable directions for further development, it is hoped that this treatise can provide a focus for this development and contribute to photovoltaics becoming an inexpensive, large-scale source of clean, high-grade energy worldwide.

References

Conibeer G, Jiang, CW, Green, MA, Harder N and Straub A (2002), "Selective energy tunnel junction contacts for potential application to hot carrier PV cells", paper submitted to WCPEC3, Osaka.

Appendix A

Greek Alphabet

Letter	Lowercase	Uppercase	Letter	Lowercase	Uppercase
Alpha	α	A	Nu	ν	N
Beta	β	B	Xi	ζ	Ξ
Gamma	γ	Γ	Omicron	o	O
Delta	δ	Δ	Pi	π	Π
Epsilon	ε	E	Rho	ρ	P
Zeta	ξ	Z	Sigma	σ	Σ
Eta	η	H	Tau	τ	T
Theta	θ	Θ	Upsilon	υ	Y
Iota	ι	I	Phi	ϕ	Φ
Kappa	κ	K	Chi	χ	X
Lambda	λ	Λ	Psi	ψ	Ψ
Mu	μ	M	Omega	ω	Ω

Appendix B

Physical Constants

Symbol	Definition	Value*
q	electronic charge	$1.602176462(63) \times 10^{-19}$C
m_o	electron rest mass	$9.10938188(72) \times 10^{-28}$g $9.10938188(72) \times 10^{-31}$ kg
π	circle circumference to diameter (calculable)	$3.14159265358979...$
c	velocity of light in vacuum (exact value)	$2.99792458 \times 10^{10}$ cm/s 2.99792458×10^{8} m/s
ε_o	permittivity of vacuum (exact value)	$8.854187817... \times 10^{-14}$ F/cm $8.854187817... \times 10^{-12}$ F/m
h	Planck constant	$6.62606876(52) \times 10^{-27}$ erg-s $6.62606876(52) \times 10^{-34}$ J-s
$\hbar$	reduced Planck constant $(h/2\pi)$	$1.054571596(82) \times 10^{-27}$ erg-s $1.054571596(82) \times 10^{-34}$ J-s
k	Boltzmann constant	$1.3806503(24) \times 10^{-16}$ erg/K $1.3806503(24) \times 10^{-23}$ J/K
σ	Stefan-Boltzmann constant $[2\pi^5 k^4/(15h^3 c^2)]$	$5.670400(40) \times 10^{-12}$ Wcm^{-2}K^{-4} $5.670400(40) \times 10^{-8}$ Wm^{-2}K^{-4}
kT/q	thermal voltage	$0.025852027(44)$ V (at 300K) $0.025692606(44)$ V (at 25°C)
λ_{1eV}	wavelength in vacuum for 1 eV photon (hc/q)	$1.23984186(11)$ μm
$SR_{1\mu m}$	ideal spectral responsivity for 1 μm photon $(q\lambda/hc)$	$0.806554477(70)$ A/W

Reference: P.J. Mohr and B.N. Taylor, J. Phys. Chem. Ref. Data, Vol. 28, pp. 1713-1852, 1999 (digits in parentheses are the estimated one-standard-deviation uncertainty in the two last digits of the given value).

Appendix C

Fermi-Dirac and Bose-Einstein Integrals

C.1 Functional Expressions

Fermi-Dirac integrals of the form

$$F_j(\eta) = \frac{1}{\Gamma(j+1)} \int_0^\infty \frac{E^j dE}{exp(E-\eta)+1} \qquad \text{(C.1)}$$

occur widely in semiconductor transport theory and carrier density calculations, where E is the energy above the edge of the respective band normalised by the thermal energy kT and η is a similarly normalised value of the carrier fermi-energy or electrochemical potential. $\Gamma(n)$ is the Gamma function equal to $(n = 1)!$ for positive integral values of n and $\sqrt{\pi}(2m-1)!!/2^m$ for $n = (m + \frac{1}{2})$, where the $!!$ sign represents the product $1.3 \dots (2m-1)$.

In a similar way, the Bose-Einstein integral is defined:

$$\beta_j(\eta) = \frac{1}{\Gamma(j+1)} \int_0^\infty \frac{E^j dE}{exp(E-\eta)-1} \qquad \text{(C.2)}$$

his integral has similar application to the case of bosons and its properties have been explored elsewhere, although not as fully as the Fermi-Dirac integrals. In the theory of optoelectronic devices such as light emitting diodes and solar cells, it has been pointed out that a non-zero electrochemical potential can be assigned to photons generated in a biased device, making the properties for non-zero arguments of increased interest. These Bose-Einstein integrals share many properties in common with Fermi-Dirac integrals explored in the present Appendix.

Generally,

$$I_j^{\pm}(\eta) = \frac{1}{\Gamma(j+1)} \int_0^\infty \frac{E^j dE}{exp(E-\eta)\pm 1} \qquad \text{(C.3)}$$

where I_j^+ represents the Fermi-Dirac integral and I_j^- represents the Bose-Einstein.

C.2 General Properties

Both integrals share the differentiation formula:

$$\frac{d}{d\eta} I_j^{\pm}(\eta) = I_{j-1}^{\pm}(\eta)$$

(C.4)

For η negative or zero, the expansion for $(1 \pm x)^{-1}$ can be used to derive the following series expression:

$$I_j^{\pm}(\eta) = \sum_{r=1}^{\infty} \frac{(\mp 1)^{r+1} \exp(r\eta)}{r^{j+1}}, \eta \leq 0$$

(C.5)

This results in the "classical" or completely non-degenerate expression:

$$I_j^{\pm}(\eta) \approx \exp(\eta), \eta \ll 0$$

(C.6)

Equation (C.5) can be used to derive the following relationship that is also valid for positive η:

$$I_j^+(\eta) = I_j^-(\eta) - I_j^-(2\eta)/2^j$$

(C.7)

This leads to a recurrence relationship:

$$I_j^-(\eta) = \sum 2^{-jr} I_j^+(2^r\eta)$$

(C.8)

More complex relationships exist between integrals of the same order but with positive and negative arguments. For example (Blackmore xxxx),

$$I_j^+(\eta) = \cos(j\pi) I_j^+(-\eta) + \frac{\eta^{j+1}}{\Gamma(j+2)} [1 + R_j(\eta)], n > 0$$

(C.9)

where $R_j(\eta)$ is a generally infinite series in negative powers of η. When j is an integer, this expression is simplified.

C.3 Special Cases

C.3.1 $\eta = 0$

Integrals involving zero argument can be expressed in terms of the Reimann zeta function $\xi(n)$ since:

$$I_j^-(0) = \xi(j+1) = \sum_{n=1}^{\infty} 1/n^j \qquad (C.10)$$

as may be seen from Eq. (C.5). Invoking Eq. (C.8), it follows:

$$I_j^+(0) = \xi(j+1)(1-2^{-j}) \qquad (C.11)$$

Since $\xi(n)$ equals $\pi^2 6$, $\pi^4 90$, $\pi^6/945$, $\pi^8/9450$ and so on for even values of n (2, 4, 6 and 8, respectively), this expression is simplified for odd integral values of j.

C.3.2 Integral j

In this case, relationships such as given by Eq. (C.9) is limited to a finite number of terms. Other simplifications occur as discussed later. The case for $j = 0$ is the highest order given analytically:

$$I_0^\pm(\eta) = \pm ln[1 \pm exp(\eta)] \qquad (C.12)$$

C.4 Approximate Expressions

Considerable effort has been invested in deriving approximate expressions for Fermi-Dirac integrals as reviewed elsewhere (Blakemore 1982).

One compact approximation for negative arguments is based on the following expression:

$$I_j^\pm(\eta) \approx \frac{exp(\eta)}{1 \pm C_j^\pm exp(\eta)} \qquad (C.13)$$

where the constant $C_j^\pm$ is chosen to minimise error over the desired range. This expression has a similar expansion to that of the integrals of interest with an identical first term. The second term is also identical if $C_j^\pm$ is given the value $2^{-(j+1)}$. $C_j^\pm$ chosen in the range:

$$2^{-(j+1)} \diamondsuit C_j^\pm \diamondsuit \pm(1/I_j^\pm(0)-1) \qquad (C.14)$$

will give the best results for negative arguments, where the value on the right is ensures that the expression gives the correct value for $\eta = 0$. The value of the left hand side will cause the expression to underestimate the integral while the value on the right will always overestimate it.

The expression is accurate over all orders with $j > 0$ to better than about 1% accuracy, and considerably better for $\eta < -1$. For the inverse problem, that of calculating η given $I_j^{\pm}$, the expression leads to:

$$\eta = ln(1/I_j^{\pm} \mp C_j^{\pm}) \tag{C.15}$$

which is accurate to better than $0.01\ kT$ for η negative.

For negative arguments, the integrals can be evaluated to any predetermined level of accuracy by using only a finite number of terms of the series expansion of Eq. (C.5). By bounding the error in the remaining terms, it follows that:

$$I_j^{\pm}(\eta) = \sum_{r=1}^{m-1} \frac{(\mp 1)^{r+1}\ exp(r\eta)}{r^{j+1}} + \Delta^{\pm} \tag{C.16}$$

$$\Delta^{\pm} = \frac{(\mp 1)^{m+1}\ exp(m\eta)}{m^{j+1}[1 \pm [m/(m+1)]^{j+1}\ exp(\eta)]} \tag{C.17}$$

$$\left|Relative\ Error\right| < \Delta^{\pm}\ exp(\eta) \tag{C.18}$$

Values of m of unity, so that only the Δ term is evaluated, are accurate to much better than 1% accuracy for $\eta < -4$. More terms are required for higher η values. Working through the series until the magnitude of the last term evaluated is consistent with the desired relative error and dividing this by the factor in square brackets before adding to the accumulated sum would be an appropriate procedure.

C.5 More General Integrals

A generalised form of the Fermi-Dirac and Bose-Einstein integrals is also of interest.

$$I_j^{\pm}(\eta, \varepsilon) = \frac{1}{\Gamma(j+1)} \int_{\varepsilon}^{\infty} \frac{E^j dE}{exp(E - \eta) \pm 1} \tag{C.19}$$

For $(\varepsilon - \eta) \geq 0$, this can be expanded in a series expansion similar to Eq. (C.5).

$$I_j^{\pm}(\eta, \varepsilon) = \sum_{r=1}^{\infty} (\mp 1)^{r+1} exp\,[\,r(\eta - \varepsilon)]\left(\mp \frac{\varepsilon^j}{r} + \frac{j\varepsilon^{j-1}}{r^2} \mp \frac{j(j-1)\varepsilon^{j-2}}{r^3} + ...\right)/\Gamma(j+1)$$

$$\tag{C.20}$$

For integral j, this has a finite $(J + 1)$ number of non-zero terms and the integral simplifies to:

$$I_j^\pm(\eta, \varepsilon) = \sum_{k=0}^{j} (\mp\varepsilon)^{j-k} I_k^\pm(\eta - \varepsilon)/(j - k)! \tag{C.21}$$

For example,

$$\beta_3(\eta, \varepsilon) = \varepsilon^3 \beta_0(\eta - \varepsilon)/6 + \varepsilon^2 \beta_1(\eta - \varepsilon)/2 + \varepsilon\beta_2(\eta - \varepsilon) + \beta_3(\eta - \varepsilon) \tag{C.22}$$

where each term on the right can be evaluated using Eq. (C.16).

The inverse of such a function for integral j can be found by expanding $I_k^\pm(\eta)$ in a power series in $I_0^\pm(\eta)$:

$$I_k^\pm(\eta) = I_0^\pm(\eta) + \sum_{r=2}^{\infty} \alpha_{k,r}^\pm [I_0^\pm(\eta)]^r \tag{C.23}$$

where

$$\alpha_{k,2}^\pm = (2^k - 1)/2^{k+1} \tag{C.24}$$

$$\alpha_{k,3}^\pm = (3^k - 1)/3^{k+1} - (2^k - 1)/2^{k+1} \tag{C.25}$$

with terms for larger values of r becoming increasingly complex. This expression shows that, for negative x:

$$I_0^+(\eta) < I_k^+(\eta) < I_0^+(\eta) + [I_0^+(\eta)]^2(2^k - 1)/2^{k+1} \tag{C.26}$$

and

$$I_0^-(\eta) > I_k^-(\eta) > I_0^-(\eta) - [I_0^-(\eta)]^2(2^k - 1)/2^{k+1} \tag{C.27}$$

Retaining only the first term in such an expansion, Eq. (C.22) can be written as a quadratic in $\beta_0(\eta-\varepsilon)$. solving for β_0 and invoking Eq. (C.12) allows η to be evaluated knowing $\beta_j(\eta, \varepsilon)$.

The actual value of η will be bounded by this value and the value found with only the first term β_0 retained, as per Eq. (C.27).

An even more general form of the integrals of interest is given by:

$$I_j^\pm(\eta, \varepsilon 1, \varepsilon 2) = \frac{1}{\Gamma(j+1)} \int_{\varepsilon_1}^{\varepsilon_2} \frac{E_j dE}{\exp(E - \eta) \pm 1} \tag{C.28}$$

The following expression can be derived from Eq. (C.21):

$$I_j^{\pm}(\eta, \varepsilon_1, \varepsilon_2) = \sum_{\ell=1}^{2} \sum_{k=0}^{j} (-1)^{\ell+1} (\mp \varepsilon)_\ell^{j-k} I_k^{\pm}(\eta - \varepsilon_\ell) / (j-k)! \qquad (C.29)$$

For example,

$$\beta_3(\eta, \varepsilon_1, \varepsilon_2) = \varepsilon_1^3 \beta_0(\eta - \varepsilon_1) / 6 - \varepsilon_2^3 \beta_0(\eta - \varepsilon_2) / 6 + \varepsilon_1^2 \beta_1(\eta - \varepsilon_1) / 2$$
$$- \varepsilon_2^2 \beta_1(\eta - \varepsilon_2) / 2 + \varepsilon_1 \beta_2(\eta - \varepsilon_1) - \varepsilon_2 \beta_2(\eta - \varepsilon_2) + \beta_3(\eta - \varepsilon_1) - \beta_3(\eta - \varepsilon_2)$$
$$(C.30)$$

An approach to finding the inverse in this case may be to note the relationship:

$$\beta_0(\eta - \varepsilon_2) = \beta_0(\eta - \varepsilon_1) - \ln[1 - \exp(\varepsilon_1 - \varepsilon_2) \beta - 1(\eta - \varepsilon_1)] \qquad (C.31)$$

or find a similar way of relating the β_0 terms with different arguments.

References

Blakemore JS (1982), Approximations for Fermi-Dirac integrals especially the function $F_{1/2}(\eta)$ used to describe electron density in a semiconductor, Solid-State Electronics 25: 1067.

Blackmore JS (1987), Semiconductor Statistics, Dover, New York.

Appendix D

List Of Symbols

α	absorption coefficient
ε	dielectric constant; normalized energy; energy
ξ	Reiman zeta function
β	Bose-Einstein integral; constant
Γ	Gamma function
λ	wavelength
μ	chemical potential; mobility
η	efficiency
ρ	density of states
σ	Stefan-Boltzmann constant; conductivity
τ	lifetime
Ω	solid angle; phase volume
A	cross-sectional area
c	velocity of light in vacuum
D	diffusion coefficient; density of states
e	base of natural logarithms
E	energy
E_c, E_v	energies of conduction- and valence-band edges
E_F	Fermi level
f	frequency; occupancy function; fraction of hemisphere
FF	solar cell fill factor
g	degeneracy
G	generation rate of electron-hole pairs per unit volume
h	Planck's constant
I	current; integral
I_o	diode saturation current
I_{sc}	short-circuit current
J	current density
J_e, J_h	electron and hole current densities
k	Boltzmann's constant; momentum
m_0	electronic rest mass
n	electron concentration; refractive index
n_i	intrinsic concentration

N_C, N_V	effective densities of states in conduction and valence bands
$\dot{N}$	photon flux
p	hole concentration
P	power; probability
q	electron charge
Q	heat
r	position vector
R	radiation per unit solid angle and per unit area
S	entropy
t	time
T	temperature
u	energy per unit volume
U	net recombination rate per unit volume
V	voltage; volume
V_{oc}	open-circuit voltage
W	work
x,y,z	position co-ordinates

Appendix E

Quasi-Fermi Levels

E.1 Introduction

The concept of a fermi level was introduced in Chap. 4 to describe the energy distribution of electrons and holes in thermal equilibrium. In non-equilibrium, quasi-fermi levels or imrefs provide a useful tool for semiconductor device analysis as outlined below. Formally, they correspond to the electrochemical potentials of non-equilibrium thermodynamics.

E.2 Thermal Equilibrium

A system in thermal equilibrium has a single, constant-valued fermi level. The case of p-n junction in thermal equilibrium is shown in Fig. E.1. For non-degenerate conditions *(n « N_C, p « N_V)*:

$$n = N_C \exp \left[(E_F - E_C)/kT \right] \qquad \text{(E.1)}$$

$$p = N_V \exp \left[(E_V - E_F)/kT \right] \qquad \text{(E.2)}$$

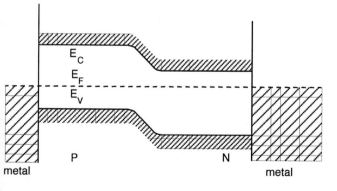

Fig. E.1: P-N junction in thermal equilibrium.

E.3 Non-Equilibrium

When voltage is applied and/or the p-n junction is illuminated, the concept of a fermi level no longer applies. However, "quasi-fermi" levels allow the analysis of semiconductors away from equilibrium. Separate electron and hole quasi-fermi levels, E_{FN} and E_{FP}, are defined by the following equations [again for non-degenerate conditions, see Eq. (5.28) of text for the more general form]:

$$n = N_C \, exp[(E_{FN} - E_C)/ kT] \tag{E.3}$$

$$p = N_V \, exp[(E_V - E_{FP})/ kT] \tag{E.4}$$

$$np = N_C N_V \, exp[(-E_G + E_{FN} - E_{FP})/ kT \tag{E.5}$$

Note that:

$$np = n_i^2 \, exp[(E_{FN} - E_{FP})/ kT] \tag{E.6}$$

The latter equation suggests that quasi-fermi levels are a measure of the level of disturbance from equilibrium ($np = n_i^2$ at equilibrium). Also, from Eqns. (5.29) and (5.30) of the text:

$$J_n = \mu_n n \frac{dE_{FN}}{d} \tag{E.7}$$

$$J_p = \mu_p p \frac{dE_{FP}}{dx} \tag{E.8}$$

Currents are determined by the product of carrier concentration and the quasi-fermi level gradients – a second useful attribute. As an aside, note that substituting Eq. (E.3) into Eq. (E.7) gives:

$$J_n = \mu_n n \frac{d[E_C + kT \, ln(n / N_C)]}{dx} \tag{E.9}$$

$$i.e., J_n = \mu_n n \frac{dE_C}{dx} + kT \mu_n \frac{dn}{dx} \tag{E.10}$$

The first term is just the drift component of current since the electric field, F, is given by:

$$F = \frac{1}{q} \frac{dE_C}{dx}$$

while the second term is the diffusion component (noting Einstein's relationship, $D_n = \mu_n kT/q$). Hence:

$$J_n = q\mu n_n F + qD_n \frac{dn}{dx} \qquad \text{(E.11)}$$

E.4 Interfaces

At interfaces between different materials, quasi-fermi levels can be discontinuous as shown in Fig. E.2 for a metal/semiconductor interface. The electron and hole currents across the interface depend on the corresponding carrier concentration near the interface and the size of the discontinuity between E_{Fm} and the corresponding quasi-fermi level, E_{FN} or E_{FP}.

For the above case of contact to p-type material (Fig. E.2), the hole concentration near the interface is large. Quite small differences between E_{Fm} and E_{FP} will support large current flows, i.e., E_{FP} would be closely aligned to the metal fermi level, E_{Fm}. The extend of the misalignment between E_{FN} and E_{Fm} for the contact to p-type material depends on the surface recombination velocity of the interface. High surface recombination velocity will enforce conditions similar to thermal equilibrium at the interface, i.e., $E_{Fm} \approx E_{FN} \approx E_{FP}$. Low surface recombination velocity will allow large separation between E_{FN} and E_{FP} with $dE_{FN}/dx \approx 0$ near the interface, for p-type material.

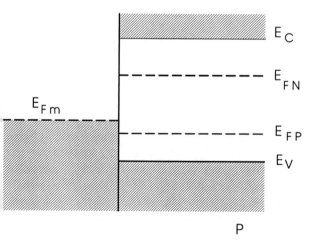

Fig. E.2: Idealized interfacial conditions for the case where both electrons and holes flow from semiconductor to metal.

E.5 Non-Equilibrium P-N Junction

If a voltage, V_a, is applied to a p-n junction in the dark, the energy-band diagram will take the form of Fig. E.3. Applying a voltage between the metal contacts gives an energy separation between the quasi-fermi levels in the two contacts equal to qV_a as shown. Across the interfaces, the majority carrier quasi-fermi levels are essentially continuous with the respective metal fermi level, while the minority carrier quasi-fermi levels need not be continuous.

In the bulk quasi-neutral regions, the majority carrier quasi-fermi levels will be approximately constant. This is because the majority carrier concentrations are large here and small quasi-fermi level gradients will support large current flows. Combined with the constancy across the metal interface, it follows that E_{FN} - E_{FP} $\approx qV_a$ in the junction region.

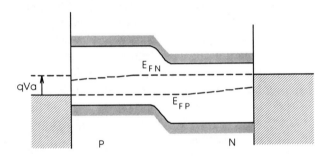

Fig. E.3: Quasi-fermi levels in a non-equilibrium p-n junction.

Hence, near the junction:

$$np = n_i^2 \exp(qV_a/kT) \tag{E.12}$$

This is an alternative derivation from that in more elementary texts (e.g., Martin A. Green, "SOLAR CELLS: Operating Principles, Technology and System Applications", Prentice-Hall, New Jersey, 1982). The assumptions required in the derivation are (i) no change in majority carrier quasi-fermi level across the contact/semiconductor interface; and (ii) constancy of quasi-fermi levels across the depletion region. In a more elementary derivation, the presentation usually manages to "side-step" the first assumption. The second is equivalent to the argument that drift and diffusion components are both large and opposing in the depletion region, with nett current flow being due to a small imbalance.

The quasi-fermi level formulation gives a procedure for checking the validity of these assumptions and refining initial estimates. Assuming constancy, first estimates of carrier concentrations and current flows could be calculated. Then, the current density Eqs. (E.7) and (E.8) could be used to calculate the variation of the quasi-fermi levels and hence allow the initial solutions to be refined.

E.6 Use Of Quasi-Fermi Levels

Quasi-fermi levels are a valuable tool in analyzing semiconductor devices. They allow energy band diagrams in non-equilibrium situations to be sketched which can then form the basis for more detailed analysis. Attributes and key properties include:

(i)   Contact regions easily handled. Differences between metal and semi-conductor quasi-fermi levels provide the "driving force" for carrier trans-port across the contact. The majority carrier quasi-fermi level is usually aligned to the metal fermi level (a discontinuity between these two levels corresponds to "contact resistance").

(ii)   Quasi-fermi levels vary only slowly with position in the semiconductor with no discontinuities except at interfaces.

(iii)   In bulk quasi-neutral regions in low injection, the majority carrier quasi-fermi level is separated from the band edge by nearly the same energy as in equilibrium. Electric fields in these regions will produce a gradient in the band edge $(qF = dE_C \cdot v/dx)$ which will largely transfer to the majority imref.

(iv)   Imrefs are approximately constant across depletion regions.

As an example of their use, an energy band diagram of a one-dimensional (1-D) model of a p-n-p bipolar transistor will be constructed with a positive bias on the first junction and a negative on the second.

Step 1:

Sketch in imrefs where values are known, namely at the contacts as shown in Fig. E.4. (Only two contacts can be shown on a 1-D model. The centre contact is modelled by forcing the majority carrier imref to a known value at a point in the n-type base region).

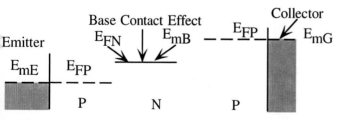

Fig. E.4: Majority carrier quasi-fermi levels at the contacts of a p-n-p transistor structure.

Step 2:

Join up the quasi-fermi levels as shown in Fig. E.5 noting that the minority and majority carriers will tend to come together at the emitter and collector contacts, if these are assumed to have a high recombination velocity (also at the base contact, although this contact is not really shown in the above diagram - it is some distance either out of, or into, the page).

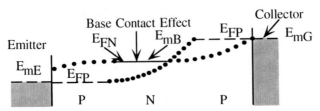

Fig. E.5: Linking quasi-fermi levels.

Step 3:

Add the energy band diagram in quasi-neutral regions as shown in Fig. E.6, noting bands lie in the same relationship to majority carrier imrefs as in equilibrium.

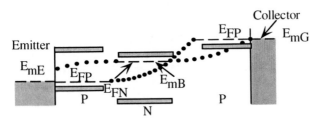

Fig. E.6: Addition of conduction and valence band edges.

Step 4:

Complete the energy band diagram by joining up bands across depletion regions as shown in Fig. E.7.

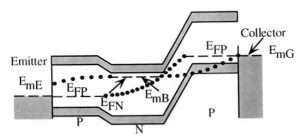

Fig. E.7: Linking of conduction and valence band edges.

Step 5:

Analyze the device. Several properties are already apparent from the above figure (e.g., $np \approx n_i^2 exp \ [(E_{mB} - E_{mE})/kT]$ near the emitter junction and $np \approx n_i^2$ exp $[(E_{mB} - E_{mC})/kT]$ near the collector junction, J_p is likely to have same direction in emitter and collector, and so on). The extension to less familiar devices such as 4-layer structures is straightforward. High injection causes the majority carrier imref to move closer to the edge of the majority carrier band than in equilibrium. However, the relative insensitivity of imref energy to such effects allow sketches even in this case. Graded bandgap problems, and more complicated contact models, are also quite easily handled.

Appendix F

Solutions to Selected Problems

Exercise 2.1

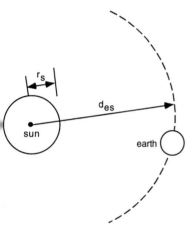

(a)   Energy at earth's mean distance/unit area/unit time
$= (r_s/d_{es})^2$ energy at sun's surface/unit area/unit time
$= (r_s/d_{es})^2 \sigma T_s^4$
Projected area of earth $= \pi r_e^2$
Surface area of earth $= 4\pi r_e^2$

Incoming energy = outgoing energy
$(r_s/d_{es})^2 \sigma T_s^4 \pi r_e^2 = \sigma T_e^4 4\pi r_e^2$
$T_e = T_s (r_s/d_{es})^{1/2}/2^{1/2} = 289.38$ K

(b)   Trick Question!

The earth absorbs 0.7 of incident radiation but also only emits 0.7 as much.
Hence the same radiation balance is maintained and:

$$T_e = 289.38K$$

(c)   The incoming radiation decreases to 0.7 but outgoing decreases to 0.6. Therefore the earth is hotter with such emission properties:

$$T_e' = (0.7/0.6)^{1/4}T_e$$
$$= 300.75 \ K$$

(d)   This increase, 2°C, is roughly that predicted by a doubling of atmospheric CO_2! The new value of emissivity is:

$$302.75 = (0.7/\varepsilon')^{1/4}T_e$$
$$\varepsilon' = 0.584$$

A 2-3% reduction is therefore required.

Exercise 3.1

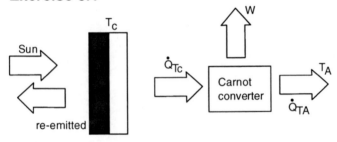

Energy from sun = $f_s \sigma T_s^4$/unit area

Energy re-radiated/unit time = σT_c^4/unit area
Energy available/unit time/unit area = $f_s \sigma T_s^4 - \sigma T_c^4$

Assume this available energy is converted at Carnot efficiency by a machine operating between a source at T_c and a sink at T_A. Possible solar energy conversion efficiency is therefore:

$$\eta = [(1-T_c^4 / (f_s T_s^4))](1-T_A / T_c)$$

Solving iteratively:

For $f_s = 1$ (full concentration), T_c must lie between T_A (300 K) and T_S (6000 K)

Try 3000 K, $\eta_{3000} = 84.38\%$
2000 K, $\eta_{2000} = 83.95\%$
2500 K, $\eta_{2500} = 85.35\%$
....
2544 K, $\eta_{2544} = 85.36\%$

For $f_s = 2.1646 \times 10^{-5}$ (diffuse sunlight), T_c must lie between T_A (300 K) and $(f_s)^{1/4}$ T_s (409 K)

Try 350 K, $\eta_{350} = 6.64\%$
    380 K, $\eta_{380} = 5.40\%$
    360 K, $\eta_{360} = 6.69\%$
    ....
    356 K, $\eta_{356} = 6.72\%$

Exercise 4.1

(Quite difficult!).

The total rate of radiative recombination occurring in a cell can be found by integrating the Shockley-Van Roosbroeck Eq. (4.8) across the cell volume (thickness W). Assuming constant quasi-Fermi level separations:

$$R_T = U_T = (WA)(e^{qV/kT} - 1)8\pi c(kT/ch)^3 \int_0^\infty \frac{n^2 \alpha \varepsilon^2 d\varepsilon}{e^\varepsilon - 1}$$

Under the conditions of the problem where α is zero for $\varepsilon \leq E_G/kT$, this simplifies to:

$$R_T = (WA)(e^{qV/kT} - 1)8\pi c(kT/ch)^3 n^2 \alpha_0 \int_{\varepsilon_g}^\infty \frac{\varepsilon^2 d\varepsilon}{e^\varepsilon - 1}$$

The rate of photon emission, at least for thick cells can be found by the Shockley-Queisser approach that leads to Eqs. (3.9) and (3.10). This is an upper bound for thinner cells, so:

$$R_E \leq \frac{2\pi(kT)^3}{h^3 c^2} A(e^{qV/kT} - 1) \int_{\varepsilon_g}^\infty \frac{\varepsilon^2 d\varepsilon}{e^\varepsilon - 1}$$

The difference is due to the photons that are recycled, i.e., generated but not emitted. This fraction is therefore given by:

$$f = (R_T - R_E)/R_T = (1 - R_E/R_T)$$

$$\geq [1 - \frac{2\pi(kT)^3}{(h^3 c^2)} \bullet \frac{(ch)^3}{W(8\pi c)(kT)^3(n^2\alpha_0)}]$$

$$\geq (1 - \frac{1}{4Wn^2\alpha_0})$$

Almost all are recycled if $W >> (1/4n^2\alpha_0)$. The problem was not intended to be any more difficult than this but some students have explored why negative values were predicted for small W.

These found $f \geq (1 - 1/2n^2)$, about 0.95. So, even when W is small, the fraction recycled stays large. This is due to the fact that photons generated by spontaneous radiative recombination are emitted in all directions, whereas Snell's law only allows those emitted in a direction close to the normal to the cell's surface to escape.

From the latter result, it can be deduced that the first result only provides a reasonably tight bound when $W >> 1/(2n^2\alpha_0)$.

An interesting follow-on problem that appears solvable is whether a cell of finite thickness has a higher efficiency than the infinite cell thickness case. See (Parrott 1993) and (Araujo and Marti 1994) where finite thickness is shown to have a definite advantage in two specific cases.

Exercise 4.3

(a) The "sky" radiation can be neglected since it corresponds to thermal generation balanced by a corresponding recombination rate at thermal equilibrium. Hence:

$$J_{sc} = qf_s \dot{N}(E_g, \infty, 0, T_s)$$

$$= qf_s \frac{2\pi(kT)^3}{h^3 c^2} \int_{e_g}^{\infty} \frac{\varepsilon^2 d\varepsilon}{e^{\varepsilon} - 1}$$

$$= qf_s \frac{2\pi(kT)^3}{h^3 c^2} \Gamma(3)\beta_2(0, \varepsilon_g)$$

where $\varepsilon_g = 1.124$ eV$/kT_s = 2.1739$. From Eq. (C21):

$$\beta_2(0, \varepsilon_g) = \varepsilon_g^2 \beta_0(-\varepsilon_g)/2 + \varepsilon_g\beta_1(-\varepsilon_g) + \beta_2(-\varepsilon_g)$$

From Eq. (C16):

$$\beta_j(\eta) = exp(\eta) + \frac{exp(2\eta)}{2^{j+1}[1-(2/3)^{j+1} exp(\eta)]}$$

with relative error equal to the second term times exp (-2.1739), i.e., 0.1137.

$\beta_0(-\varepsilon_g) = 0.1137 + 0.0067$
$\qquad = 0.1207$ (rel. error ~ 0.0007)

$\beta_1(-\varepsilon_g) = 0.1137 + 0.0034$
$\qquad = 0.1171$ (rel. error ~ 0.0004)

$\beta_2 (-\varepsilon_g) = 0.1137 + 0.0017$
$= 0.1154$ (rel. error ~ 0.0002)

$\therefore \beta_2 (0, \varepsilon_g) = 0.6552$ (rel. error < 0.0001)

$J_{sc} = 2,869$ A/cm^2 ($f_s = 1$)
$= 62.1$ mA/cm^2 ($f_s = 2.1646 \times 10^{-5}$)

(b)   In the Shockley-Queisser approach:

$$J_0 = f_c \frac{2\pi(kT)^3}{h^3 c^2} \int_{\varepsilon_g}^{\infty} \frac{\varepsilon^2 d\varepsilon}{e^\varepsilon - 1}$$

In this case, $\varepsilon_g = 1.124 \ eV/kT_c = 43.478$

Hence, $\beta_0 (-\varepsilon_g) \approx \beta_1 (-\varepsilon_g) \approx \beta_2 (-\varepsilon_g) \approx exp (-\varepsilon_g)$

$V_{oc} = (kT/q) \ ln \ [(J_{sc} + J_0)/J_0]$
$= 889.4 \ mV \ (f_c = 1, f_s = 2.1646 \times 10^{-5})$
$= 1,167.1 \ mV \ (f_c = 1, f_s = 1)$

Note that the latter is larger than the bandgap! This means that the formulation is invalid in this region since non-degenerate carrier populations have been assumed. Also, once quasi-Fermi levels move into the bands, electron states at the bottom of the conduction band become more occupied than those at the top of the valence band. Stimulated emission increases and the effective absorption coefficient becomes negative.

For an infinitely thick cell, this prevents the progression of quasi-Fermi levels into bands. For thin cells, however, penetration can occur in principle (Parrott 1986).

Chemical Potential

At open-circuit, $J = 0$ so Eq. (4.19) becomes:

$$f_c \dot{N}(E_G, \infty, qV_{oc}, T_c) = f_s \dot{N}(E_G, \infty, 0, T_s) + (f_c - f_s) \dot{N}(E_G, \infty, 0, T_c)$$

$$\beta_2(qV_{oc} / kT_c, E_G / kT_c) = \frac{f_s T_s^3}{f_c T_c^3} \beta_2(0, E_G / kT_s) + \frac{(f_c - f_s)}{f_c} \beta_2(0, E_G / kT_c)$$
$$= 5,242 \ (f_c = f_s)$$
$$= 0.1135 \ (f_s = 2.1646 \times 10^{-5}, f_c = 1)$$

Now,

$$\beta_2(\eta,\varepsilon) = \varepsilon^2 \beta_0(\eta-\varepsilon)/2 + \varepsilon\beta_1(\eta-\varepsilon) + \beta_2(\eta-\varepsilon)$$

Note that β_1 and β_2 are bounded as $\varepsilon \to \eta$ but $\beta_0 \to \infty$. This will limit ε to a value less than η. Neglecting β_1 and β_2 terms gives:

$$\beta_0(\eta-\varepsilon) < 5.546 \, (f_c = f_s)$$

From Eq. (C.12), $1 - exp\,(\eta - \varepsilon) > e^{-5.546}$
$$\eta - \varepsilon < - 0.0039$$

$V_{oc} < E_G/q - 0.00010 \ V$

Similarly, assuming $\beta_1 = \beta_2 = \beta_0$ gives:

$$\beta_0\,(\eta - \varepsilon) > 5.297$$

This gives:
$$0.00010 \ V < (E_G/q - V_{oc}) < 0.00013 \ V \ (f_c = f_s)$$

i.e., $V_{oc} = 1.1239 \ mV$.

Following the same procedure in the case where $f_s \ll f_c$ gives:

$$0.2334 < (E_G/q - V_{oc}) < 0.2346$$

A tighter lower bound in this case can be obtained by assuming $\beta_1 = \beta_2 = \beta_0 = exp$ $(\eta - \varepsilon)$.

This gives:
$$0.2346 < (Eq/q - V_{oc}) < 0.2346$$

i.e., $V_{oc} = 889.4 \ mV$.

This is a similar result to that from to the Shockley-Queisser approach.

As the two theories are more similar for smaller voltages, they will be more nearly similar near the maximum power voltage than near open-circuit. Hence, the Shockley-Queisser theory, obeying the traditional solar cell law but having a higher V_{oc}, will have a lower fill factor.

(c)   At 300 suns, the same approach gives similar values in both cases:

$$V_{oc} \ (Shockley\text{-}Queisser) = 1.1049 \ V$$
$$V_{oc} \ (Chemical \ Potential) = 1.0364 \ V$$

Exercise 5.1

A close-up of 5.2 is shown in Fig. F.1.

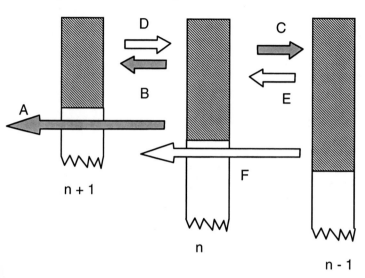

Fig. F.1: Close-up of central cells in Fig. 5.2 showing the radiatively emitted current components.

As well as the currents shown in Fig. F.1, we have the most important current, that generated by sunlight and the background thermal radiation:

$$S = qA[\, f_s \dot{N}(\, E_{Gn}, E_{G(n+1)}, 0, T_s\,) + \dot{N}(\, E_{Gn}, E_{G(n+1)}, 0, T_A\,)]$$

The current-voltage relationship for cell n is therefore given by:

$$I_n = S - (A + B + C) + D + E$$

where

$$(A + B + C) = 2qAf_c\dot{N}(\, E_{Gn}, \infty, qV_n, T_c\,)$$
$$D = qAf_c\dot{N}(\, E_{G(n+1)}, \infty, qV_{n+1}, T_c\,)$$
$$E = qAf_c\dot{N}(\, E_{Gn}, \infty, qV_{n-1}, T_c\,)$$

Note that, in the case on a monolithic tandem cell (all fabricated on one piece of semiconductor material of varying composition), components B, D, C and E would all be enhanced by a factor of n^2, where n is the refractive index. Rather than the performance limit in this case being a few percentage points below that of the filtered tandem cell limits of Table 5.1, it will lie a few percentage units below.

Exercise 5.2

The optimum operating voltage for each cell is given by the solution of Eq. (5.3). The easiest way to solve it is probably just, in each case, insert the given value of $E = E_G$, and vary V_m between 0 and E_G/q until the value giving zero is found.
Following this procedure gives:

(a) $E_G = 0.25\ eV$, $V_m = 6.0\ mV$
(b) $E_G = 0.50\ eV$, $V_m = 158.9\ mV$
(c) $E_G = 2.00\ eV$, $V_m = 1.517\ V$

These points should lie on the μ_{opt} versus E line for the diffuse case shown in Figs. 6.8 and 7.3.
The reason these voltages are so small for the small values of E is that the light coming from the sun is very weak at these energies for the diffuse case, while the light emitting ability of the cell is comparatively high. Only a small chemical potential enhancement is feasible, otherwise the cell will emit more light than it absorbs.

Exercise 6.1

The efficiency of a hot carrier cell is given by:

$$\eta = P_{use}\ /\ Solar\ Input$$

where P_{use} is given by Eq. (6.8).

This becomes:

$$\eta = \frac{\begin{aligned}&[\ f_s\dot{N}(\ E_G,\infty,0,T_s\ )+(\ f_c - f_s\ )\dot{N}(\ E_G,\infty,0,T_c\ )- f_c\dot{N}(\ E_G,\infty,\Delta\mu_H,T_H\ )]\Delta\mu_H T_c/T_H \\ &+[\ f_s\dot{E}(\ E_G,\infty,0,T_s\ )+(\ f_c - f_s\ )\dot{E}(\ E_G,\infty,0,T_c\ )- f_c\dot{E}(\ E_G,\infty,\Delta\mu_H,T_H\ )](1-T_c)\end{aligned}}{f_s\dot{E}(\ 0,\infty,0,T_s\ )}$$

For the case of $E_G = 0$ and $\Delta\mu_H$ a large negative, the corresponding terms become:

$$\dot{N}(\ 0,\infty,0,T\ ) = \frac{2\pi(\ kT\ )^3}{h^3 c^2}\xi(\ 3\ )\Gamma(\ 3\ )$$

$$\dot{E}(\ 0,\infty,0,T\ ) = \frac{2\pi(\ kT\ )^4}{h^3 c^2}\xi(\ 4\ )\Gamma(\ 4\ )$$

$$\dot{N}(\ 0,\infty,\Delta\mu_H,T\ ) = \frac{2\pi(\ kT_H\ )^3}{h^3 c^2}\int_0^\infty \frac{\varepsilon^2 d\varepsilon}{e^{\varepsilon - \Delta\mu H/kT_H} - 1}$$

When $\Delta\mu_H$ is a large negative:

$$\dot{N}(0,\infty,\Delta\mu_H,T_H) = \frac{2\pi(kT_H)^3}{h^3c^2} e^{\Delta\mu_H/kT_H} \int_0^\infty \frac{\varepsilon^2 d\varepsilon}{e^\varepsilon}$$

where the latter integral equals 2. Similarly:

$$\dot{E}(0,\infty,\Delta\mu_H,T_H) = \frac{2\pi(kT_H)4}{h^3c^2} e^{\Delta\mu_H/kT_H} .6$$

Substituting these into the expression for efficiency gives:

$$\frac{\Delta\mu_H(T_c/T_H)(1/T_s)}{\Gamma(4)\xi(4)/[\Gamma(3)\xi(3)]\}}\left[1+\left(\frac{f_c-f_s}{f_s}\right)\left(\frac{T_c}{T_s}\right)^3 - \frac{(f_c/f_s)}{\Gamma(3)\xi(3)/2}\left(\frac{T_H}{T_s}\right)^3 exp(\Delta\mu_H/kT_H)\right]$$

$$+\left(1-\frac{T_c}{T_H}\right)\left[1+\frac{(f_c-f_s)}{f_s}\left(\frac{T_c}{T_s}\right)^4 - \frac{(f_c/f_s)}{\Gamma(4)\xi(4)/6}\left(\frac{T_H}{T_s}\right)^4 exp(\Delta\mu_H/kT_H)\right]$$

Equation (6.12) follows from this.

Substituting numerical values and finding optimal values of $\Delta\mu_H$ gives:

$$T_H = 4000K, f_s = 1, \Delta\mu_H = -0.7485eV, \eta = 86.7\%$$
$$T_H = 3600K, f_s = 2.1646x10^{-5}, \Delta\mu_H = -3.1738eV, \eta = -22.8\%!!$$

The negative value is a bit of a surprise since earlier calculations (Ross and Nozic 1982), although not calculating for $E_G = 0$, suggest values close to 68% should be obtained.

Exercise 7.1

For $mE_G \le E \le (m + 1) E_G$, $\mu(E) = mqV$, where m is the integral part of E/E_G. As $E_G \to 0$:

$$\mu(E) \approx qV (E/E_G) = aE/E_G$$

If we define a temperature T_l such that:

$$(E - aE/E_G)/kT_C = E/kT_l,$$

the analysis will be greatly simplified. Rearranging

$$a/E_G = (1 - T_C/T_I)$$

Substituting into Eq. (7.3)

$$P = (1 - T_C/T_I)[\,f_s\dot{E}(0,\infty,0,T_s) + (f_c - f_s)\dot{E}(0,\infty,0,T_A) - f_c\dot{E}(0,\infty,0,T_I)\,]$$

Dividing by $f_s\dot{E}(0,\infty,0,T_s)$ gives:

$$\eta = (1 - T_C/T_I)[\,1 + (f_c - f_s)(T_A/T_s)^4/f_s - f_c(T_I/T_s)^4/f_s\,]$$

Which equals Eq. (7.4).

Note that something a little weird has happened here. One would expect a/E_G to approach infinity as $E_G \rightarrow 0$. However, the approach to zero must be in such a way that mE_G remains finite to satisfy the constraints on the problem.

Exercise 8.1

(a)

$$(I_{NM}/A)_{max} = q[\,f_s\dot{N}(E_A,E_B,0,T_s) + (f_c - f_s)\dot{N}(E_A,E_B,0,T_A) - f_C\dot{N}(E_A,E_B,0,T$$

For I_{21}, parameters are:

$E_A = 0.7$ eV, $E_B = 1.2$ eV, $T_s = 6000$ K, $T_C = T_A = 300$ K, $f_c = 1$, $f_s = 2.1646 \times 10^{-5}$

Giving $(I_{21}/A)_{max} = 28.9$ mA/cm^2

For I_{32}, changed parameters are:

$E_A = 1.2$ eV, $E_B = 1.9$ eV

Giving $(I_{32}/A)_{max} = 30.2$ mA/cm^2

For I_{31}, changed values are:

$E_A = 1.9$ eV, $E_B = 3.1$ eV

Giving $(I_{31}/A)_{max} = 21.8$ mA/cm^2

(b) To find variation with chemical potential assuming the Shockley-Queisser approximation, a term given below is subtracted:

$$(I_{NM} / A)_{dark} = f_c \dot{N}(E_A, E_B, 0, T_c)(e^{\mu_{NM} / kT_c} - 1)$$

This gives standard solar cell current-voltage type curves with open-circuit values of 0.47 eV, 0.94 eV and 1.61 eV.

(c) Curves for I_{21} and I_{32} are combined using the standard approach, e.g., Fig. 5.5 (Wenham et al. 1994) since there is a mismatch in the current, the curve will be a bit flattened from a normal solar cell curve with a short-circuit current density of 28.9 mA/cm^2 and an open-circuit potential of 1.41 eV.

(d) The curve from (c) is combined with the I_{31} curve as in Fig. 5.2 of the above reference to give the final output curve, which again looks fairly similar to a single solar cell curve (no humps or bumps!) with a short-circuit voltage of 1.41 eV.

(e) From the curve (d), the maximum power output of 62.8 mW/cm^2 at a potential of 1.276 eV is found, compared to the solar input calculated as 159.1 mW/cm^2, corresponding to an efficiency of 39.5%.

References

Araujo GL and Marti A (1994), Absolute limiting efficiencies for photovoltaic energy conversion, Solar Energy Materials and Solar Cells 33: 213-240.

Parrott JE (1993), Radiative recombination and photon recycling in photovoltaic solar cells, Solar Energy Materials and Solar Cells 30: 221-231.

Parrott JE (1986), Self-consistent detailed balance treatment of the solar cell, IEE Proceedings 133: 314-318.

Wenham SR, Green MA and Watt ME (1994), Applied Photovoltaics, Bridge Printers, Sydney.

Index

Printing and Binding: AZ Druck und Datentechnik GmbH, Kempten

Printed in the United Kingdom
by Lightning Source UK Ltd.
127182UK00002B/46-51/A